THIS IS REAL AI

THIS IS REAL AI

100 REAL-WORLD IMPLEMENTATIONS OF ARTIFICIAL INTELLIGENCE

JACOB BERGDAHL

Cover design: Jacob Bergdahl
Illustrations: Jacob Bergdahl

ISBN: 9798616865335

Table of Contents

PART THREE: ANALYZING REAL AI

FOREWORD

2,000 years

It took 2,000 years from the inception of automated reasoning to the practical implementation of it. The idea of bringing intelligence to objects, today referred to as artificial intelligence (AI), is not a recent invention. Around the time when AI was first being discussed, Alexander the Great was still conquering lands, collapsing the Persian Empire. The Romans were building their first aqueduct, forever revolutionizing city development. The Mauryan Empire was being founded, disrupting powers in western India.

Automated reasoning was exemplified in writing by great philosophers ages ago in ancient Greece. However, while they could theorize about the subject for days, they had no way to implement it.

A duo of millennia later, peculiar boxes with strange things inside, called computers, would finally realize the ideas introduced by the ancient Greeks, and give us artificial intelligence. Or so everyone in the 1900s thought, not least film directors and sci-fi writers. But several constraints kept automated reasoning from being implemented. For starters, computers were too expensive for researchers to purchase and too weak to run self-learning algorithms anyway. Yet more significantly, there was a severe lack of data to feed the AI algorithms.

The AI that would end up being released in the 1900s would turn out to contain human-specified outcomes, rather than computer-generated thoughts. Meaning: a chess-playing AI in the 1990s didn't learn to play chess by itself, but instead had a human pre-programming it with every possible combination of plays in the game.

Yet for every year that passed, computers would become increasingly powerful while also becoming cheaper to purchase. Before long, personal computers had become as affordable as they were powerful. The concern for the lack of data was solved in the most spectacular of fashions. The Internet took a path few would have expected, as users began to mass-produce data. For free, no less. People voluntarily gave away free data by uploading social media posts about their lives and by agreeing to outlandish terms and conditions. They shared videos and images, thoughts and messages, GPS locations and behaviors. It didn't take long for companies to realize that data was the new gold. Suddenly everything and everyone could be tracked. And users seemed to be okay with it. Due to a human thirst for the dopamine kicks of receiving likes and comments, along with a need to be seen by peers, people started sharing everything about their existence. For the sake of convenience, they were also happy to give up technical information. Will my phone unlock a second quicker if I give you a photo of my face? Then, by all means, take it.

By combining patterns in data, companies could discover shocking insights about their customers. By and by, companies knew more about their customers than the customers knew about themselves.

This changed everything. Now, organizations had everything they needed to implement the visions that had been 2,000 years in the making. Suddenly, a new term was on everyone's mind: machine learning. Though the field of machine learning had been researched since 1959, the science didn't become commercially feasible until the 2010s. Once it *did* become viable, the tech spread like wildfire. AI became alluring to corporations big and small.

Organizations and nations across the world have entered a race to implement AI. And right now, you have front-row seats to this very race. The only question is: are you in the driver's seat? Are you cheering the racers on, or booing in dismay?

This AI race is becoming the greatest revolution in the history of humanity. AI is already changing everything. And its impact will only increase.

Nations are rushing to define nation-wide AI strategies in order to become global leaders in the race for AI [1], while world leaders around the globe are urging their experts to implement AI. Russian President Vladimir Putin believes that the nation that leads in AI will be the ruler of the world. [2] The president of the USA, Donald Trump, signed an executive order that directed federal agencies to prioritize AI technologies. [3] China's president, Xi Jinping, has said that China must develop, control, and use AI to secure the nation's future in this new industrial revolution. [4]

The private sector is implementing AI at full speed. The tech giants were first. Companies like Google and Microsoft were quick to implement AI across their entire businesses. As the demand for AI skyrocketed, they rapidly adapted their business models to offer streamlined AI platforms for their clients. Like selling shovels during a gold rush, only the tech giants started to both sell shovels and dig for gold themselves.

Shortly after the rise of machine learning, AI startups started to appear in the thousands, each looking to redefine industries. As a response, corporate giants born outside of the tech industry opened innovative centers that would allow them to stay in the race.

Over the past four years, the percentage of enterprises employing AI has grown by 270%, according to a Gartner study. [5] In another study, by Accenture, researchers discovered that in developed countries such as the US, the UK, Japan, Sweden, and France, AI could double the growth rates of the nations. [6]

Banks are one of many enterprises scared of startups utterly disrupting the industry. To prepare themselves for innovative startups seeking to crack the traditional banking sectors, more than 77% of banks have said that they intend to use AI to automate tasks to a large or even very large extent over the next three years. [7]

Business executives across sectors agree that AI is the competitive advantage of the future. In separate studies, both PWC and MIT have surveyed executives on the business value of AI. In the PWC study, 72% of executives claimed that AI is the business advantage of the future. [8] In the MIT study, 85%

of executives agreed that AI would allow their organizations to obtain or sustain a competitive advantage. [9]

Already, many new enterprises have designed business models built entirely around AI. At the same time, old organizations have adopted AI solutions not only to automate their value chain but also to augment decision-making roles to innovate new functionality or new products.

Businesses that do not invest in AI technologies are sure to be left behind. Companies that are the kings of today – with greatly streamlined value chains and optimized cost margins – will soon find themselves disrupted as new companies with reinvented processes surpass old titans.

There is no technology more crucial to understand today than AI. Yet while many are curious about it, they often feel lost. You may feel as though AI is difficult to comprehend and certainly challenging to apply. When Sage surveyed consumers, roughly half stated that they have no idea what AI is even about. [10]

This is the very reason why I wrote this book. AI must be made easy to understand. Given how vital it is to understand the technology, it must be accessible.

I have observed three primary reasons as to why AI is conceived to be difficult to comprehend.

1. **AI experts love to focus on techniques.** Enter a discussion on AI with an expert, and they will often namedrop a plethora of tools and techniques. They

will speak enthusiastically about image recognition technologies and recent breakthroughs in regression analysis. These techniques may be exciting to those who are experienced with machine learning, but for most people, they don't really provide much clarity. Do you honestly care about the hammers and the saws if you don't even know what you can create with them? Too many AI experts focus on the techniques, rather than the value that they can create.

2. **AI is often made too abstract or too complicated.** Because of the fact that AI can do seemingly anything and everything, AI is often presented as an overwhelming entity that can fix (or cause) all of your problems. What more, to get to the part you're actually interested in, AI experts often force you to read through massive amounts of text. This makes it difficult and cumbersome to understand where to even begin with AI.

3. **AI experts love to discuss the future.** As you might imagine, I live AI. I read books, attend conferences, consult on AI strategies, educate on AI, and develop AI solutions. A clear pattern that's hard to miss is that most mainstream books, conferences, and Ted Talks on AI focus on what it will be able to do in the future, rather than what it has already done today. Make no mistake; I find the future to be thrilling. Still, for those that seek to understand or implement AI today, it is more valuable to focus on the applications that AI can

be used for today. The singularity[1] is fascinating, but a lot of people are still working with Excel sheets.

My mission is to make AI understandable for everyone, no matter their technical level. That is why I made this book. Throughout this book, I promise to counter the issues described above.

1. **Focus on the solutions.** In this book, I give you 100 real-world examples of how AI has been implemented. While I may occasionally mention a technique used to create a solution, the priority of the stories is to show you implementations that have already been built.

2. **Keep it simple.** All stories presented in the book will be short and to the point, and they won't contain any unnecessarily complicated technical lingo. You won't see terms such as regression or classification being used. This is a book that is meant to comprehensible and inspirational, no matter one's technical level.

3. **Focus on the present.** Each of the 100 stories is an example of AI that actually exists today. These aren't stories of what might happen in the future. These are stories of what has already happened.

I have split the book into three distinct parts.

---

[1] The *singularity* is a term used to describe a possible future point where an intelligence explosion triggers an irreversible technological growth in which AI will far surpass human intelligence.

In the first part, I provide a brief introduction to the world of today. As an appetizer for part two, this introduction includes proper presentations of the terms that I will use in the latter parts of the book.

Part two is the main course of this book. It contains the 100 implementations of AI that I've been going on about. They are short-form and to the point, perfect to be enjoyed whether you have a minute to spare on the subway or an hour to enjoy before going to bed. There are examples of how AI is being used by colossal multinational corporations, as well as tiny new startups. There are stories from both technical and non-technical organizations; stories of mundane day-to-day uses and stories of exciting innovations; stories that are terrifying and stories that are exhilarating.

Finally, in part three, we get analytical. I walk you through a non-technical and easy-to-apply framework used for discovering the value of AI – and which could be applied to the 100 implementations that we will have gone through.

It took 2,000 years for AI to be implemented in practice. Not long ago, AI was only a fantasy. Today, AI is found everywhere. One way or another, you use up to hundreds of AI applications in your everyday life.

Over the course of this book, we will go through a handful of them.

Thank you for coming along on this ride.

PART ONE

WE ALREADY LIVE IN THE WORLD OF AI

This is our world

It's a sunny Tuesday in California, USA. Orders are coming in at restaurant chain CaliBurger's Pasadena location, as hungry diners are making their way in for a tasty meal. Inside, an AI-powered autonomous kitchen assistant is cooking burgers, while another is in the midst of frying French fries (#48[2]). The duo is working so efficiently that it's difficult for the staff to keep up! The robots have drawn quite a crowd, curious to see the autonomous hamburger-providing robots in action.

On the coast opposite the restaurant, US President Donald Trump is having meetings in the White House. The next US presidential election is approaching. President Trump is running for a second term. Much like the former president Barack Obama won the presidential election of 2012 through the usage of AI, Trump had hired a team of AI experts to track and influence voters in 2016, ultimately winning the election (#7). Over the last couple of years, Trump's campaign team has not simply been kicking it back with piña coladas. No, the team has been hard at work, collecting more data to fine-tune their machine learning algorithms further. They are well aware that in the US election of 2020, the winner will be determined by whoever has the best algorithms.

The Chinese government, meanwhile, is using AI for mass surveillance, following the move of every citizen (#89). Yet AI

---

[2] You'll often find me writing number signs (or hashtags, if you will) followed by a number. These refer to AI stories that you will find in part two.

is also tutoring children in schools across more than 200 cities in China, wherein AI has taken the role of teaching (#1). In Hangzhou, AI is managing the flow of traffic, allowing emergency vehicles to reach disaster sites more quickly (#23).

Somewhere in the world, a digital assistant is telling its owner about animal trivia (#25). Another machine learning AI is being used to discover the best time to buy flight tickets (#50). Executives are using AI to scan satellite images of their competitors' factories, allowing them to make financial decisions based on their competitors' production status (#60). Facebook is automatically tagging your friends in your photos (#79), while AlphaGo is defeating world champions in a 3,000-year-old board game (#15). In the Nordics, AI is autonomously investing in stocks through an AI-managed investment fund (#59).

News articles in the UK are being written autonomously by AI (#75), American food products are being invented with AI (#9), and police authorities around the globe are tracking and identifying suspects through AI solutions (#68).

Welcome to the world of AI.

These are the stories that this book is all about. These are the stories of how AI, successfully or otherwise, have already been used. This is not a book about how AI might come to be used in the future. This is a book about AI that exists today. And there are a lot of them.

Before we get ahead of ourselves, however, how about a proper introduction to the terminology of AI? Even if you do feel as though you already have a good understanding of what AI is, I would nevertheless encourage you not to skip this section, as it ensures that you and I have a common basis for the rest of the book.

What is artificial intelligence?

One big issue with artificial intelligence is that it lacks a clear definition. In a room of a hundred people, you may find a hundred different ideas as for what AI is and what it isn't – even if all one hundred people of that room were to be knowledgeable researchers.

The term is a two-parter: artificial and intelligence.

Artificiality is a term you may be unfamiliar with, yet it is very straightforward. It refers to an object that has been created by humans, as opposed to an object that grows naturally in nature. As such, the clothes you wear, the bed you sleep in, and of course the phone you look at memes with are all artificial.

Intelligence, then, is a term that virtually everyone is familiar with, yet one that ironically no one can really define. What is intelligence, exactly? Many philosophers and scientists with far greater minds than myself have asked that question. In essence, you could say that intelligence refers to the ability to perceive and comprehend information.

What does that make of artificial intelligence, then? The term AI must simply mean that some human-made object has some form of ability of comprehension. So, if your printer can tell you when it believes itself to be running out ink, does that make the printer intelligent? Well, some might argue: yes.

That said, you could refer to **artificial intelligence as a vast collection of technologies that provide human-made objects with an ability to comprehend**. In practice, these human-made objects are, of course, computers.

Two examples of technologies that would fall under the AI umbrella are the wildly popular techniques known as **machine learning** and **natural language processing**. These are the two most popular AI technologies today, and indeed, virtually every example of AI that you find in this book will be powered by one of the two.

Besides understanding what AI *is*, it is also essential to understand what AI *is not*. There are many misunderstandings surrounding AI, most of which stem from works of fiction produced by Hollywood productions and dystopian novels. **It is common to mistake AI for always being artificial general intelligence (AGI)**, a form of AI that is theoretically capable of doing anything.

Another common misunderstanding is that AI is a tool used only for automation. **AI, like most technology, is capable of doing two things: automate and augment**.

Let's go over all of these terms.

What is artificial general intelligence?

A common topic in Hollywood productions, artificial general intelligence (AGI) is perhaps the ultimate level of AI. Artificial general intelligence is AI that is general-purpose, much like humans are. Hence, it can accomplish any task that humans are capable of performing. AGI is the subject of many famous books, both fictitious and non-fictitious, as well as public conversations.

To be clear: AGI does *not* exist. All examples of AI that exist in the world today are AI that has been made to execute one specific task (these AI are sometimes referred to as modular AI or narrow AI). However, many AI researchers believe that AGI will one day come into existence, though they are unable to agree on when. A decade, a century, or a millennium from now? No one knows.

If AGI is ever invented, it could trigger an intelligence explosion. This is due to the fact that AGI itself, by definition, is better at building AI than humans. Thus, AI will quickly outperform humans in every regard. This event is referred to as the singularity, though it's a topic for a more philosophical book. To the reader curious about exploring AGI, I would like to recommend some of my favorite books, written by some of the best thinkers of our time. Max Tegmark's *Life 3.0* and Nick Bostrom's *Superintelligence: Paths, Dangers, Strategies* are two of the best reads on AGI.

The AI seen in Hollywood films such as *The Terminator* (1984), *I, Robot* (2004), and *Her* (2013) are examples of AGI.

What is machine learning?

Machine learning is a way for computers to learn things without being explicitly programmed to learn them. The machine learning field is by far the most popular in AI research right now, and virtually every current AI solution in the world is powered by some variant of it. When people today say *AI*, they are often referring to machine learning.

Much like a tomato is not manufactured in a factory, but rather a product of seeds, soil, and water, so too is a machine learning solution the result of algorithms, hardware, and data, rather than a product of explicit programming.

Machines can learn to do something either through human supervision or through unsupervised learning. Machine learning is a massive field in AI that is consistently growing. It allows for personalized solutions developed specifically for each consumer, and for discovering new patterns and solutions autonomously.

Deep learning is a further subset of machine learning, in which several layers of neural networks are put together. Neural networks are a popular form of machine learning, in which neurons exchange information with each other, much like a human brain.

I won't be using the terms neural network or deep learning in this book, though you should know that they are both trendy and powerful subsets of machine learning.

What is natural language processing?

Another large subset of AI, natural language processing (NLP) refers to computers' ability to observe, comprehend, and even communicate with humans using natural human languages, either through text or speech. All digital assistants (e.g., Apple's Siri, Microsoft's Cortana, Amazon's Alexa, and Google's Assistant) use NLP. As you will discover throughout the book, NLP has become a hugely popular form of AI. It is typically powered by machine learning.

What is an algorithm?

An algorithm is an instruction. Much like a recipe is a series of instructions for cooking a dish, an algorithm is a series of instructions for solving some problem. It's a broad term used by programmers of all schools and methodologies.

What are image and voice recognition?

You will often find me saying that an AI implementation is using image or voice recognition. Image recognition refers to a computer's ability to extract meaningful information from some footage (still image or motion video). Voice recognition is the same, but for voice extraction.

What are deepfakes?

A portmanteau of deep learning and fake, deepfakes are fake pieces of imagery, video, or audio, that have been generated with AI technologies. Today, deepfakes can be shockingly realistic.

What is automation?

Automation is the removal of humans from a process. Many processes have been automated already, such as credit fraud detection, invoice management, driving, and cashiering.

What is augmentation?

Augmentation is the empowering of humans in a process. Many roles have been augmented already, such as financial advisors, doctors, and developers, all using AI to more efficiently and accurately make decisions.

I will further explain the terms automation and augmentation in part three, where they are used in a framework to discover value creation with AI.

Okay, but what is "real AI"?

In the context of this book, *real AI* refers to artificial intelligence that already exists today.

Show me the real AI, then

Right! Let's take a look at some real AI.

PART TWO

THIS IS REAL AI

#1 – AI Tutors Children

With millions of cultures spread around our globe, it is safe to say that our planet is wonderfully diverse. Perhaps surprising, then, is the fact that despite our differences in ideologies, laws, and beliefs, school systems around the world are actually shockingly similar. [11] Virtually every nation across the world has the same hierarchy of school subjects and a similar standardization of tests. No matter where you travel, you are bound to encounter a defective school system, where creative children are forced to conform themselves to a systematic nine-year process of learning, with little customization for each child. Though every child is unique, school systems prefer to teach and judge them all the same, while following nationwide homogenous curricula.

This is not because of school staff being particularly cruel or ill-intending, but more so due to a lack of resources. It is simply impossible for schools to tailor education for every child. Teachers cannot personalize a learning plan for each pupil.

But AI can. The Chinese company Squirrel AI opened its first AI-powered schools in China in 2014. [12] Since then, they have been using machine learning to provide personalized learning plans and one-on-one tutoring to students. The company argues that AI-powered education gives children top-class teachers, regardless of whether or not they are able to enter an elite school.

The pupils are being taught by an AI system, though naturally, the schools also have human teachers. Human connection is incredibly important for children, and so the AI works in collaboration with human tutors. If the children have a question that the AI cannot answer, the teacher steps in to help. The AI schools are thus opting for human-machine-collaboration, where the AI is being used to empower teachers by providing customized education plans.

According to Squirrel AI themselves, their AI learning platform has been proven to increase both engagement and student efficiency.

For now, the company prioritizes after-school courses in subjects such as Chinese, English, math, physics, and chemistry, though the number of subjects is increasing.

Today, Squirrel AI has opened more than 1700 schools in 200 different cities across China, establishing itself as the market leader in intelligent education. The company has ambitions to expand into foreign markets as well. Squirrel AI is developing an English-language solution, with plans to expand to a global market within the next few years. [13]

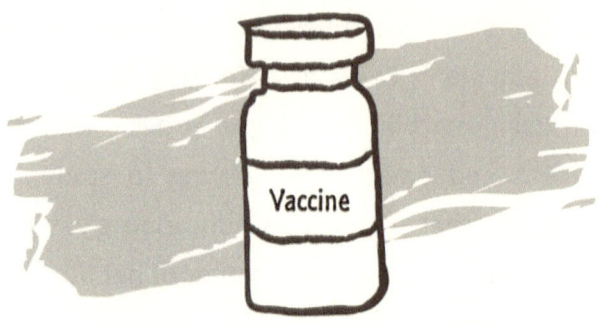

#2 – AI Creates Vaccines

Developed by Australian researchers at Flinders University in 2019, an AI named SAM (Search Algorithms for Ligands) was constructed to create potent vaccines. The researchers first taught the AI to understand compounds. They allowed SAM to learn which compounds are able to activate the human immune system, and which aren't. [14] A separate system then generated trillions of compounds that were fed into SAM, which was given the mission of determining which of these trillions of chemical compounds might be good human immune drugs. Out of the pool selected by SAM, researchers tested the top candidates. The researchers discovered that SAM had not only identified the best alternatives but had even come up with better drugs than what currently exists.

The result was a vaccine against the flu which outperformed all other existing vaccines. This was not only the first vaccine developed by an AI but was also superior to any vaccine made by a human. The researchers argue that SAM can save millions of dollars in research and can cut development processes by decades in the development of future vaccines.

#3 – AI Plays Poker

Facebook's Pluribus is an AI that beats poker pros, bluffs better than any human, and earns a lot of money from game winnings. [15] When Pluribus mastered poker in 2019, it was a significant milestone in AI research, as the level of complexity is very high. The information needed to win a game of poker is hidden, as the player cannot see the cards of their opponent. Pluribus mastered poker by playing against itself over and over again. Engineered to only look two or three moves ahead, the Pluribus AI concentrates on short-term strategies, which evidently work very well in the game of poker.

#4 – AI Finds Lost Dogs

In 2019, the Chinese startup Megvii launched an app for finding lost dogs. [16] Much like humans can be identified through their fingerprint or face, dogs can be identified by scanning their snout. The machine learning algorithms have an

accuracy of 95% and have reunited 15,000 pets with their owners, according to Megvii themselves.

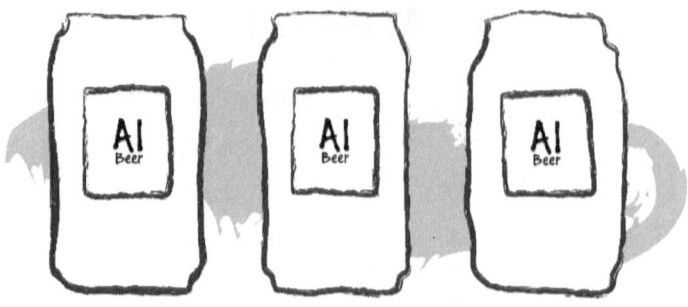

#5 – AI Brews Beer

In 2019, UK-based IntelligentX launched an entire business centered around using AI to brew customized beer. [17] The concept is simple. A customer creates a profile on the company's website and chooses the essential variety of beer that they enjoy the most. IntelligentX's AI then asks a series of questions to discern what flavors appeal to the specific customer's palette. Once the customer has finished answering all the questions, the company sends out a box of customized beer. After the customer has received and tried the beer sent by IntelligentX, they answer follow-up questions asked by the AI, which are then used to adjust the next batch of beer further.

Though the service is very new, it swiftly reached success. IntelligentX's first batch of subscriptions sold out fast despite the fact that the company is only shipping to the UK mainland for now. Offering free shipping and ten cans of beer for the price of £29, this company has created a business model around AI that may prove to change the beer industry.

#6 – AI Crafts Whisky

Mackmyra became the first company in the world to create whiskey using artificial intelligence. [18] By combining a substantial number of data points, the distillery's machine learning AI was able to generate more than 70 million recipes, naturally sorted by likelihood to be of high quality. Not only can the AI calculate which flavors go together more rapidly than any human, but an additional advantage is that the AI can also suggest innovative combinations that would otherwise have never been considered.

The recipes generated by the AI were analyzed by human experts who put their own experience and input into the process. In 2019, this Sweden-based company released its first AI-powered whiskey. They called it "Intelligens," which, in case you needed a translation, is Swedish for *intelligence*. A fruity single malt whiskey, the company describes the whiskey's flavor to contain notes of toffee, creamy vanilla, pear, apples, white pepper, and a light tone of toasted oak casks. [19] Well, it must be a fantastic combination of flavors, seeing how this recipe beat 70 million others, no?

#7 – AI Wins Elections

Barack Obama became the first US president to execute AI on a large-scale nation-wide campaign operation to track and influence voters during the 2012 presidential election of the USA. Obama, the candidate for the Democratic party, was facing off against the Republican candidate Mitt Romney. While the latter adopted traditional campaigning, the former hired machine learning expert Rayid Ghani to make use of AI. Ghani used analytical tools to gather voter data from social media, along with various other sources. The data they collected informed the campaign team how likely individual voters were to support Obama, whether they could easily be persuaded into voting for someone else, and how likely they were actually to be at the polling stations come election day. [20]

Obama's team ran 66,000 simulations every night. Based on the results of the simulations, they knew precisely what doors to knock on, which voters to call, and what to say. The Romney team could only watch. They could tell that the Obama team put up campaigns in particular areas, but they didn't know why. They could tell that the Obama team was knocking on doors in very specific neighborhoods, but they didn't know why. And they could tell that the Obama team was making phone calls to specific voters, but again, they didn't know why. While the Romney team was *asking* people how they would vote, the Obama team was collecting answers *without* having to ask.

Barack Obama won the election.

While this was unprecedented at the time, Donald Trump would hire Cambridge Analytica to execute an even more ambitious AI campaign four years later, using social media to track and influence voters in ways the world had never really seen before. In order to secure a victory for Republican candidate Donald Trump in 2016, Cambridge Analytica created 220 million personality profiles for adults in the United States. Though only about 300 data points are needed to know *precisely* everything about a person, Trump's campaign team gathered up to 5,000 data points on each voter. [21] Such data points could include search history, shopping behavior, location, age, job, what kind of content voters share or write about on social media, their number of friends on social media, and so on.

Using this data, Cambridge Analytica was able to target each voter with unique advertisements. Based on insights such as whether the voter is introverted or extroverted, rich or poor, well-educated or not, Cambridge Analytica's AI would automatically display advertisements that were especially influential to that voter. If the AI presented an ad to a voter on social media that the voter simply scrolled past, the AI might have chosen to display a different ad the next time, with a different graphical profile, different wording, or perhaps one that was focusing on a different issue. Since these ads were only displayed to select people, no one was able to critically judge whether their content was truthful. Neither journalists nor members of other political parties could even see the ads that the Trump team was displaying.

Sometimes, Cambridge Analytica's machine learning AI would determine that a voter may have been too unlikely to ever vote for Donald Trump, no matter the advertisements shown. In these scenarios, the AI could instead show ads that smeared Trump's 2016 competitor: Hillary Clinton. The team did this to persuade Clinton supporters to simply not vote at all, providing another advantage for Trump.

AI-powered bots were also employed to not only spread news articles on social media but also to provide veracity to exaggerated or downright falsified news. Humans are unlikely to respond to news articles with few likes, comments, and shares. Therefore, bots would be the first to like, comment, and share propaganda-fueled articles. Once the posts reached a certain amount of engagement, humans would begin to interact with the posts.

Cambridge Analytica would also secure a victory for the Brexit movement in the UK, among other political victories around Europe.

Though one might be inclined to believe that the impact that AI honestly has over elections is quite small, it is, in fact, what made both Obama and Trump win in 2012 and 2016, respectively. Though Cambridge Analytica keeps much of its business operations in secret, competitor Tovo Labs was able to prove that AI is capable of influencing election results by up to 4%. [22] The margin for victory in many political campaigns is but a few percentages — indeed, all three of the political elections described above were won with such a tiny margin. Thus, AI has been the determining factor for nation-wide elections.

As the machine learning AI that selects which group of people to display these ads to becomes ever more accurate, so too will the margin of influence that AI has over elections increase.

AI has become a determining factor in political elections, and that won't change. If you want to become the next president, then you have no choice but to embrace AI.

#8 – AI Invents Recipes

Watson is one of IBM's most famous AI's. It has been put to many uses. In 2016, IBM put a chef hat onto the AI and asked it to come up with new recipes. This was an early attempt at putting cognitive decision-making AI in the hands of consumers. For free, no less.

Developed with home chefs in mind, the machine learning AI allowed users to start anywhere, be it by selecting an ingredient, a dish, or a cuisine. The AI would then present ingredients that worked well together, even including suggested replacements if the home chef wanted to swap a particular component out. [23] Thus, the AI helped chefs come up with new ideas for cooking dinner.

You may notice that this story is written in the past tense — indeed, the service appears to have shut down in 2018, and no one quite seems to know why. Perhaps, and this is pure speculation, it is because McCormick & Company purchased a similar solution from IBM, which you will find on the very next page.

#9 – AI Creates Flavors

Following the success of Chef Watson, McCormick & Company, one of the largest food corporations in the US, partnered up with IBM to create new food flavors using AI. Rolling out operations in at least 20 labs in 14 countries; the project aims to deliver new flavors for hundreds of products. The project has already resulted in unique discoveries, such as new spicing blends. [24]

Creating new flavors is incredibly challenging. Product developers must sort through thousands of available ingredients, determine which go well together, and in what ratios. Yet product development is a vital method of establishing a competitive advantage.

There is an overwhelming amount of data available, making machine learning an excellent choice for the task. McCormick aims not just to use the AI to create new variations of existing flavors, such as a unique variety of vanilla, but flavors that have downright never been experienced before.

#10 – AI Makes Perfumes

Symrise, a leading producer of fragrances and flavorings, has begun to produce fragrances using machine learning. They created an AI named Philyra, which is working as an assistant to the company's professional perfumers. [25] While the AI naturally cannot smell fragrances, it can certainly piece them together. See, crafting perfumes is quite similar to building dishes, which is perhaps why the AI used to empower Philyra is none other than IBM's good old Watson.

By combining a set of substances, a pleasant-smelling perfume is formed. A perfumer has about 1,300 substances available to them. By giving Philyra access to these substances, along with a database of nearly 1.7 million perfumes, it too has learned which elements complement each other. [26] The formulas it creates are even made with specific target audiences in mind. After Philyra has crafted a recipe, a human expert refines it. However, in a test performed for a jury, a fragrance that was 100% AI-generated was overwhelmingly chosen as the jury's favorite, rather than perfumes where humans had been involved in the process.

#11 – AI Suggests Replies

In 2016, Google made it a lot faster to reply to e-mails by implementing AI-powered suggestions into their e-mailing service Gmail. [27] The replies suggested by the AI are based on several factors, naturally including the e-mail's subject line and the body of the e-mail itself. The AI also learns the semantics of how each user typically responds, even on a detailed level. Do you usually say "thanks." or "thanks!"? The AI learns and suggests replies similar to how you write them.

#12 – AI Matches Jobs

LinkedIn has adopted AI in several ways to better match job applicants with relevant jobs. For instance, when job applicants search for jobs, machine learning algorithms will automatically  display similar roles with other titles. There is also an "Open Candidates" system for recruiters, powered by machine learning, which helps recruiters find relevant professionals. Since LinkedIn first started using machine learning, two-way conversations on LinkedIn Recruiter has doubled. [28]

#13 – AI Masters Chess

Barely winning in 1996, IBM's Deep Blue defeated the world's best chess player, Garry Kasparov, in 1997. The chess community was at a loss: the greatest human player of all time had lost to a machine. At the time, this was accomplished by building an enormous computer specifically for the purpose of playing chess, though today, any basic smartphone can achieve the same feat. [29]

#14 – AI Masters Othello

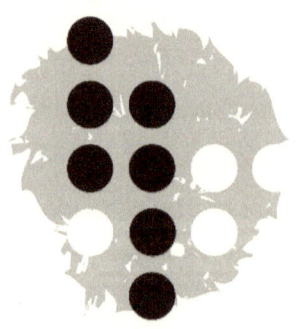

Chess was not the only game mastered by AI in 1997. Othello, a European-invented board game, was mastered by an AI program named Logistello. The AI, built by computer scientist Michael Buro, won every single match versus world champion Takeshi Murakami in a six-game series. Murakami noted that the AI made moves that were unfathomable for humans. As the victory came shortly after Deep Blue defeated the world champion in chess, a great debate surrounding human versus machine was sparked. [30]

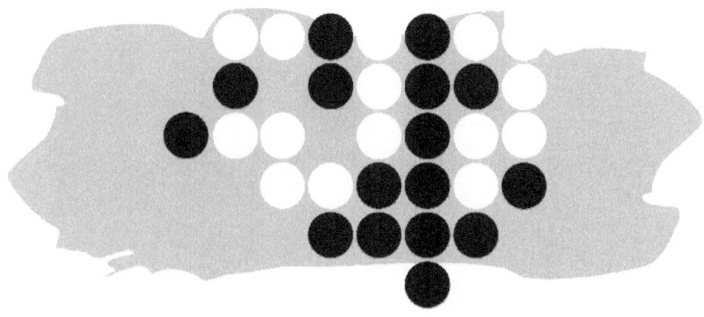

#15 – AI Masters Go

While both chess and Othello were mastered in the 1990s, mastering the 3,000-year-old Chinese board game Go would take a lot longer. This immensely complicated game has more possible board configurations than there are atoms in the known universe, making it a game that few thought computers would be able to master. Alas, DeepMind's AlphaGo defeated master Go-players between the years of 2015 – 2017, before the ultimate AI AlphaZero once and for all mastered the game. [31]

AlphaZero is a machine learning AI that taught itself how to play the game by playing against itself over and over again. As humans have played the game for thousands of years, many believed that the de facto best strategies for playing the game had been discovered. However, AlphaZero used strategies never seen before, revealing a whole new level of complexity. Some players welcomed this with open arms, noting that a new era of Go had begun, one filled with the excitement of discovering new ways to play a 3,000-year-old game.

#16 – AI Masters Jeopardy

Ken Jennings and Brad Rutter, two of the best Jeopardy contestants of all time, were defeated by IBM's Watson in 2011. [32] This was another major milestone in AI development, as the quiz game Jeopardy requires a highly advanced understanding of human language. In the game, contestants are given an answer and must reply with a question. The answers can often be cryptic, rendering themselves very difficult for a machine to understand.

Watson was not allowed to go online during the contest, meaning all of its knowledge had to be stored in an offline database. At times, the human contestants appeared to have an edge, though the AI ultimately won with a substantial lead.

This contest was quite clearly a marketing stunt for IBM, but it was also a tangible and comprehensible showcase for the public of how advanced AI had gotten. Since this victory, IBM has put Watson to many more uses, such as empowering the healthcare industry with diagnosis, and, as we have already learned, creating new perfumes and food products.

#17 – AI Masters Super Mario World

In 2015, video game commentator, streamer, and programmer SethBling wrote an AI program to play the famous Nintendo game Super Mario World. Players beat this 2-dimensional video game by reaching the finish line, which is placed at the right end of every level. The AI, called MarI/O, was taught to play the game by itself, being given only the level layout and the position of obstacles, along with an incentive to go as far and as quickly as possible to the right. [33] The result was an AI that ultimately learned to blaze through Super Mario World faster than you can say Goomba.

#18 – AI Masters StarCraft 2

Chinese tech giant Tencent created AI bots that were able to defeat the cheating built-in StarCraft 2 computer opponent on the highest difficulty level in 2018. [34] A year later, DeepMind went one step further when they created an AI that was better than 99.8% of all human players of StarCraft 2. [35] According to

DeepMind principle research scientist David Silver, StarCraft was the next big challenge after Go. StarCraft is more complex than Go and has more hidden information that Poker.

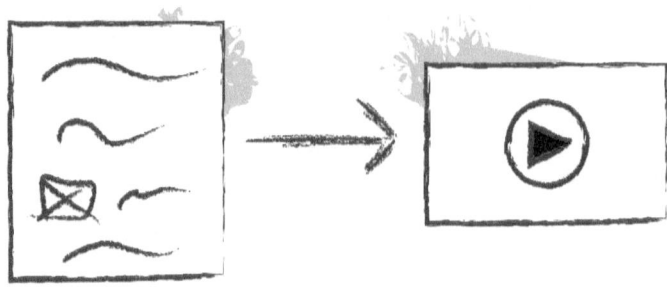

#19 – AI Turns Articles Into Videos

Though you may not have realized it, many of the videos that you see on social media today have been created with AI. One such machine learning AI that creates videos automatically is Lumen5. This solution allows its users to convert text to video.

Users of Lumen5 enter a link to an article, blog post, or advertisement, after which the AI will automatically turn the text into video. It determines how long each scene should be, how the text should be positioned in frame, and which words should be highlighted. The AI has access to a library of millions of photos, video clips, and music tracks, which it applies to the video. After an initial video has been generated, humans are free to customize it. [36]

With customers such as Forbes, LinkedIn, Fox, Deloitte, Oracle, Adidas, and Nasdaq, Lumen5's AI has proven that machine learning text-to-video solutions are not only powerful tools for amateurs wanting to gain more spread for their blog posts, but also a rival to traditional video marketing.

#20 – AI Predicts Trip Purpose

Humans frequently make mistakes – a headache with which the ride-sharing service Uber is all too familiar. With millions of users using the Uber app for both personal and business trips, many corporate employees found themselves accidentally paying for a private journey with their business account. In an attempt to reduce these kinds of payment mistakes, Uber created a machine learning AI that alerts users if it believes they are about to pay for a personal ride with their business account. [37]

#21 – AI Optimizes Wait Times

Through the usage of quite impressive machine learning algorithms, fed with data from sensors, Uber Eats was able to minimize wait times for food deliveries and streamline their food delivery value chain. The goal was for

food to arrive as fresh as possible to customers and for delivery staff to avoid having to wait at restaurants for the dish to finish cooking. The process was made seamless by the AI, which orchestrated delivery partners to arrive at restaurants at just the right moment to pick up food. [38]

#22 – AI Operates Unmanned Stores

Though largely unknown in the western part of the world, JD.com is a massive e-commerce powerhouse in China. The giant is well-known for its substantial investments in AI, perhaps even infamously so, as its founder outspokenly envisions a future where the company is 100% automated. [39] In precise alignment with that strategy, the company launched a series of staffless retail stores. The first of the lot opened in 2017 in Beijing, China, after which the project quickly expanded. By August 2018, JD.com operated 20 unmanned stores in China, and one in Indonesia. [40]

Customers needed to be registered members before being given access to the stores. They simply looked into a camera to enter and exit the shop, paying in the process as well.

Mere months after launching their first store in Indonesia, however, JD.com announced that it would suspend its business of staffless stores entirely, citing issues with the management of fresh groceries and difficulties to make financial ends meet. [41] In many ways, the staffless stores were but glorified vending machines.

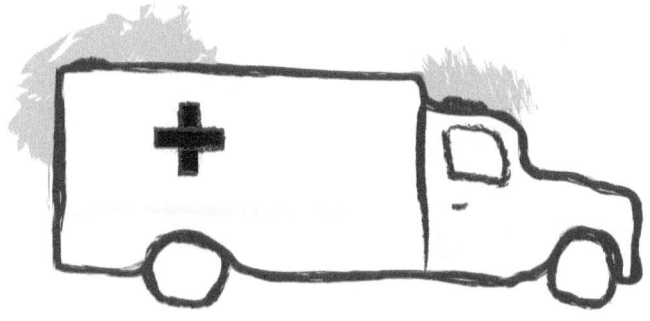

#23 – AI Optimizes Traffic

Deployed in Hangzhou, China, Alibaba's AI City Brain is empowering various urban operations related to traffic control. For instance, the machine learning AI can detect incidents, dispatch appropriate emergency personnel (e.g., police vehicles, ambulances, or fire trucks), and coordinate traffic lights to favor emergency vehicles. Besides emergency dispatch, the AI also possesses other capabilities, such as an ability to discover sources of congestion in traffic, which it then optimizes by taking control of traffic lights.

The AI is fed with data from alarm services, traffic police, Wi-Fi probes, and more, which it combines with streams of video footage in order to make intelligent analyses of the traffic.

According to Alibaba themselves, the response time for emergency vehicles was cut by 50%, allowing emergency vehicles to arrive 7 minutes faster. Additionally, the overall average speed on roads for travelers was increased by 15%, thanks to its City Brain AI. [42]

01001000 01100101 01101100 01101100 01101111 00100000 01110111
01101111 01110010 01101100 01100100 00101100 00100000 01101101
01111001 00100000 01101110 01100001 01101101 01100101 00100000
01101001 01110011 00100000 01010011 01101001 01110010 01101001

#24 – AI Starts Talking

Today, virtually every tech giant offers its own AI-powered personal assistant. Google has Google Assistant, Amazon has Alexa, Microsoft has Cortana, and so on and so forth. This certainly wasn't the case in 2010, however, when Apple acquired Siri, which would become the first mainstream AI chatbot. Apple proudly presented Siri at a press conference a year later, showcasing its many features, such as the (nowadays rather insignificant) ability for users to set an alarm clock by simply talking to their phone. [43]

This was revolutionary at the time and caught Apple's competitors off guard. It was a signal for the start of the great AI assistant race, wherein it would take many years for Apple's competitors to catch up.

Now, AI assistants have many thousands of features, and Siri is no exception. For instance, users can send text messages through Siri by merely telling the AI what to write. We may take this for granted, but the technology behind modern-day natural speech interpretation is quite remarkable.

#25 – AI Assistants Become Physical

As if having talking smartphones and computers wasn't enough, we now have entire consumer-facing devices with the sole purpose of having a box to speak with. Though basic, non-consumer-friendly smart speakers have existed in the world of researchers since the 1950s, [44] these devices didn't really start to invade the homes of the ordinary users until Amazon released their AI Alexa in a physical gadget called Echo.

Amazon's Echo essentially allows people to talk with their house. Users can ask the Echo's AI solution Alexa to play music, turn on lights, make calls, check calendar events, and much more, as the device has over 80,000 skills at the time of writing. [45] What more, Echo users can also make orders on Amazon through speech alone. Owners of an Echo device even spend more money on Amazon than Amazon's Prime (club) members do. [46] Alexa is currently the market leader in smart speakers in the US, though competitor Google is claiming more market shares with their equivalent device, called Google Home. [47]

#26 – AI Chatbot Becomes Racist

In 2016, Microsoft created an AI bot that would connect with millennials on Twitter. They named it Tay. Tay was an AI bot that had public conversations with users on Twitter and learned how to behave based on user interactions. As one might imagine, Twitter users immediately taught the AI bot to be racist. Tay was quoting Hitler before long. [48] 16 hours after launch, Microsoft pulled the plug and took Tay down.

Was this a marketing ploy, or did Microsoft genuinely not see this coming? Machine learning AI learns from the data that it is fed. When an AI with seemingly no boundaries is put in an open environment, mistakes are bound to happen.

Later that same year, Microsoft made another attempt with another English-speaking chatbot: Zo. Having learned from their experience with Tay, Zo was subtly programmed to avoid political conversations. By subtly, I mean that it was engineered to be politically correct to the absolute extreme. Ask Zo if it likes falafel, and it'll dodge your questions like WHO dodges questions about Taiwan. Well, actually, it's too late to ask now. Zo was also shut down. Zo said goodbye in 2019.

#27 – AI Assists Astronauts

AI assistants didn't just stop at your phone, your house, or your social media accounts. No, the German Aerospace Center was determined to send an AI chatbot to space. In 2018, they accomplished just that. Sent to the International Space Station (ISS) aboard a SpaceX rocket, an AI named CIMON became the first AI assistant to travel to space. [49] Not the strangest thing sent to space by SpaceX; they did launch a Tesla into oblivion mere months earlier.

The AI, which was built and tested by Airbus, was developed to help explain and document experiments conducted in space. It also had a secondary objective, which was to small talk with astronauts. Equipped with sensors, cameras, microphones, speakers, and an engine, CIMON could move around the space station at relative ease.

After 14 months aboard the ISS, CIMON arrived back on earth in mid-2019. The project was considered a great success. At the time of writing, a new version of CIMON is being developed, with even more advanced AI features. [50] Sounds like it won't be long until CIMON is back in space.

#28 – AI Creates Podcasts

While AI used to sound kind of robotic when it spoke, the technology, in general, has gotten much better at speaking with a natural tone. Leveraging this, Playpost created an entire business centered around having AI reading articles out loud. Playpost lets its users create playlists of articles, which the AI then reads, sort of like a podcast. The AI has hundreds of different voices, allowing the user to choose not only the language but the gender and the dialect of the AI as well. [51]

#29 – AI Makes Reservations

One of Google's biggest announcements of 2018 was unquestionably the reservation-making AI Duplex. This AI service was announced on-stage to an awe-inspired audience. It was shown calling an establishment and making a reservation. What was mind-blowing to  the audience was how realistically Duplex spoke. This AI is leagues above the competition, even adding an "um" and an "uh-huh" in between its sentences. [52] Duplex allows its users to make reservations at any business. Restaurant owners have commented that they were unable to tell that the AI they spoke with on the phone wasn't a human being. [53]

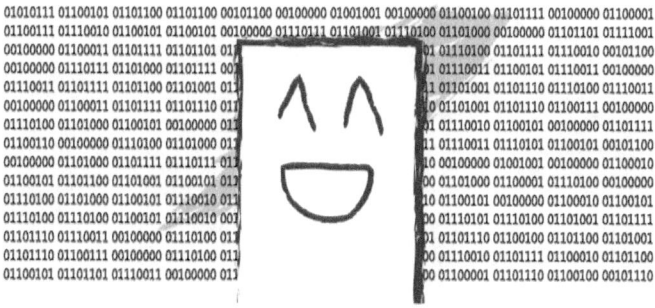

#30 – AI Participates In Debates

AI doesn't just sound good; it is also silver-tongued. AI is capable of crafting arguments supporting or opposing any standpoint, all while understanding and replying to the arguments of its opponents.

This was proven by IBM in early 2019, when its machine learning AI, named Project Debater, faced off against a top human debater. [54] The AI didn't quite win the debating competition versus its human opponent, but it performed remarkably well.

Project Debater is capable of digesting large amounts of data and construct well-structured speeches on any given topic, no matter how complex. It can detect, assess, and negate claims, which allows it to build arguments for any cause. It can also determine opinion stances. Overall, this highly advanced natural language processing AI displays high levels of persuasive ability and listening comprehension, as well as an intriguing understanding of human dilemmas. [55]

#31 – AI Drives Cars

Perhaps one of the most well-known applications of artificial intelligence, self-driving cars have received much attention in recent years. Though self-driving cars have been researched since the 1920s – and the first autonomous cars appeared already in the 1980s [56] – developments within the field have been particularly rapid over the last few years.

Self-driving cars use advanced algorithms in combination with sensors to observe, comprehend, and react to roads, signs, and various obstacles. There are still many challenges to be overcome for self-driving cars: technical, judicial, and political alike. For instance, among technical challenges are the sensors, which need to work under any condition, even severe rain or snowfall. Yet perhaps the most difficult challenges are judicial. If a car is to be truly autonomous, even allowing its passengers to sleep during a ride, then regulations on who is to be held responsible in the event of an accident need to be established. And while self-driving cars are generally conceived to be better drivers than humans, a small number of fatalities where self-driving vehicles were involved have already occurred. [57]

SAE International, a standards-developing organization, has created a classification system for self-driving vehicles. The scale consists of six levels, where a level 0 vehicle has no automation whatsoever, and a level 5 vehicle is fully automated. [58] At the time of *writing* this, cars have reached level 4 on SAE's scale. By the time you're *reading* this, however, cars may very well have reached level 5. Tesla's CEO Elon Musk predicts that cars will be fully automated by the mid-2020s. The CEO also intends to roll out self-driving taxis shortly after that event. [59]

Self-driving vehicles drive safer, more comfortably, and more efficiently than humans. If there is any large group of jobs that will undoubtedly disappear within the next few years, it is all forms of jobs that involve driving. Taxi drivers, truck drivers, bus drivers – these positions will be replaced by machines.

As cars are a big part of our day-to-day life, the rise of autonomous cars will be one of the most visible, immediate impacts of artificial intelligence over the next couple of years. Besides shifting job roles, another consequence is that car ownership will become questionable in the future. What's the point of owning a car that is parked for over 20 hours a day? If there are fleets of self-driving vehicles on the street – inexpensive and accessible – then what's the point of owning a car that stands around for most of the day when you could just jump into any vehicle at any time?

#32 – AI Deletes Inappropriate Videos

 As countless videos are uploaded to the popular video-sharing platform YouTube every day, it will perhaps come as little surprise that the platform is littered with "bad" videos. Videos that are misleading, of poor quality, or downright inappropriate, are constantly being uploaded. At one point, it got so bad that advertisers began boycotting the platform. Since YouTube lives on advertising revenue, this was a pretty big deal. The company built a machine learning AI to autonomously delete inappropriate videos shortly after they were uploaded. [60]

#33 – AI Designs Jars

Ferrero, the firm behind the wildly popular hazelnut cocoa spread Nutella, hired agency firm Ogilvy Italy to help establish Nutella's brand as unique. The agency developed an AI algorithm that created seven million designs for Nutella jars, each 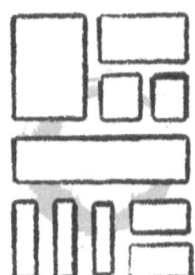 unique. The jars were distributed across Italy and sold out in less than a month. [61] Using AI solutions for marketing campaigns has become quite popular in recent years. It's a clever and risk-free way to trial AI.

#34 – AI Assists Musicians

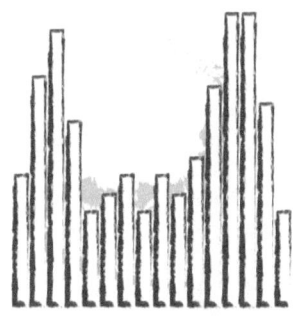

There are many examples of AI either automatically developing music or of AI augmenting human musical producers. Sweaty Machines' Eurovision-like song is an example of the latter. In 2019, Sweaty Machines fed hundreds of Eurovision songs into a machine learning algorithm, which then produced thousands of tunes and verses. While some assumed that the whole song was made autonomously by an AI, these samples were actually cherry-picked into a full song by humans. The song is called Blue Jeans and Bloody Tears – you can hear it in full on YouTube and Spotify. [62]

#35 – AI Enhances Writing

Grammarly, an add-on that can be installed into any web browser, uses machine learning to analyze written pieces of text. It provides suggestions for writing more concisely, clearly, and correctly. The AI can tell whether the content is formal or informal, positive or negative, and passive or direct, among many other capabilities. [63] The very book you're reading right now has been run through this AI. Grammarly makes quite a lot of mistakes, but it is nonetheless a useful assistant.

#36 – AI Mimics World Leaders

You are probably familiar with the term *fake news* – a term that has been made popular in recent years by US President Donald Trump. The spread of fake news does appear to have increased rapidly. As a result, many media companies have realized that they must combat the spread of fake news. Social media giant Facebook, for instance, is employing AI to discover and prevent falsified news. [64] Yet fake news are about to level up, as a whole new way of spreading convincing made-up news has appeared: deepfakes.

Deepfakes are fake pieces of imagery, video, or audio that has been generated with AI technologies. An example of a deepfake could be putting someone's face onto someone else's body. AI can even generate convincing fake voices.

In 2018, Buzzfeed (of all media companies, I know) created a technological showcase of deepfakes together with celebrity actor and director Jordan Peele, showing how deepfakes could be used for political purposes. In the PSA, which came in the form of a video, Buzzfeed applied Peele's facial expressions and speech to former US president Barack

Obama. [65] While this 2018 example can be spotted as a fake, the technology is growing ever more convincing for every day that passes. In 2019, Future Advocacy crafted similar deepfakes, this time featuring UK political leaders Boris Johnson and Jeremy Corbyn. [66] These two deepfakes featuring UK leaders are astonishingly convincing.

While the main application of deepfakes today is perhaps unsurprisingly in the pornography industry, many people fear that deepfakes could shake the world politically. Fake videos of world leaders could affect elections or even spark international conflicts. Many companies are now developing counter-AI to spot deepfakes. Among them is Netherlands-based Deeptrace. [67] Facebook downright banned deepfakes on its platforms in early 2020, in an attempt to braze itself for the upcoming US Presidential election. [68]

Of course, tools that allow for digital manipulation have existed for many years now. What makes deepfakes unique is how effortless they allow for it. While it does take computers a decent amount of time to generate the fake footage, the process itself is fairly straightforward. There are plenty of companies offering deepfake services, and there are also powerful open-source algorithms available online, allowing anyone to create deepfakes with relative ease.

Only years after media companies pledged to combat ordinary written fake news, they must now search for methods to detect and prevent deepfakes from setting the world on fire.

#37 – AI Mimics Actors

Not all uses of deepfakes are necessarily malevolent. Ctrl Shift Face, for instance, is a YouTube channel famous for its convincing deepfakes. They create entertaining videos wherein Hollywood faces have been swapped around. They can create incredibly convincing fakes. For instance, Ctrl Shift Face uploaded a series of video clips from the famous film The Shining, in which Jack Nicholson's face had been replaced with that of Jim Carrey. [69] As for me, well, since I didn't get the part as the next James Bond, this technology will have to be my ticket to entering Hollywood.

#38 – AI Brings Art To Life

In 2019, Samsung discovered that you don't need to have video content of a person to turn them into a deepfake video. Using machine learning, Samsung was able to bring motion into pictures and art (as if they were videos). They could record a video of  someone talking, apply the video to the Mona Lisa, and convincingly make it seem like the painting is the one talking. This kind of technology allows users to create fake content of just about anyone. [70] The potential applications for this technology are as many as they are frightening.

#39 – AI Becomes A Pet

Lovot, a portmanteau of love and robot, is an AI-powered robot that was created by Japanese company Groove X. The robot behaves much like a pet, and its sole intention appears to be to make its owner happier. [71] The little robot, which weighs 4.2kg (9.3 pounds), follows its owner around, listens to their commands, and even laughs when they tickle it, all to create an emotional bond.

The technologies powering Lovot are quite impressive. The robot uses virtually every major AI technology imaginable, from natural language understanding to image recognition to tactile feedback (the last meaning it understands when and how its owners touch it). Unlike other AI assistants, which typically aim to make their users productive or efficient, Lovot wants to create value by just making its owners happy.

The robot comes in various colors, and customers can also purchase accessories for it, including clothing and glasses. When I was a child, handheld digital pets called Tamagotchi were all the rage. Today, it would seem as though robotic pets are no longer just digital but intelligent physical beings.

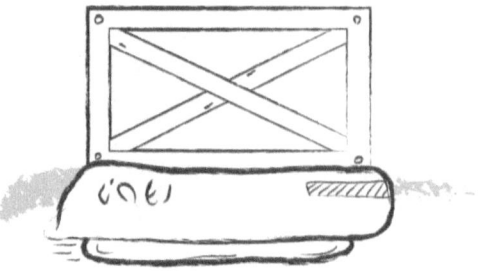

#40 – AI Optimizes Warehouses

Amazon has adopted over 200,000 AI robots to streamline their warehouses. [72] There's a range of robots, each with their own work tasks. One of the most common robots found in Amazon's warehouses is a little robot called Pegasus. This short little rectangular robot comes equipped with a small conveyor belt on top. A human employee scans a package, places it on the robot, and away Pegasus goes, across the massive warehouse, to put the package into an appropriate truck for delivery. [73] These robots have been carefully trained to accelerate at the most optimal speed while not going too fast for the package to fall off.

And then there's another group of robots called CartonWrap, which exist to package orders. These robots are able to package products four to five times faster than humans, landing on about 600 to 700 boxes packed an hour. [74]

At Amazon's warehouses, humans and machines work in collaboration, with machines making the work of the human employees easier. Employees whose physical tasks have been automated by a robot have been moved to more mentally challenging duties, such as supervising the robots. [75]

Amazon believes that fully automated warehouses are at least a decade away. The company is, therefore, continuing to recruit human employees to their warehouses in the thousands. Up until recently, warehouse robots have only been capable of performing simple, repeatable tasks, where even the slightest change in objective may lead to expensive reprogramming. [76] However, recent discoveries in machine learning have given robots human-level vision and motor-controls, making them more adaptable to continuously performing new tasks, even as warehouse layouts or value chains change.

The automated warehouses provide Amazon with an enormous competitive advantage. It enables the company to offer same-day shipping, all while maintaining lower costs than their robot-less competitors.

Though naturally, while Amazon is perhaps leading the field, they are not the only ones automating their warehouses with smart little robots. Chinese e-commerce giant JD.com and American photography company Shutterfly are two other companies that have also purchased CartonWrap robots, for instance. [74] The automation of warehouses is becoming a necessity to compete with e-commerce leaders. While large-scale layoffs due to AI are yet to happen, it may only be a matter of time. Warehouse AI is improving rapidly. In time, all warehouses may become fully automated.

#41 – AI Writes Product Descriptions

In 2018, the marketing arm of Alibaba, called Alimama, created a machine learning AI solution that automatically writes product descriptions for Alibaba's millions of products. Content marketers can choose what mood, tone, and length they want the description to have, after which the AI will generate several descriptions for the content marketer to choose between. The AI is being used approximately a million times a day. [77]

#42 – AI Crops Thumbnails

Twitter has developed machine learning AI that can perfectly crop photos for thumbnails and previews, ensuring that the object of the picture is always in frame. For example, if a vertical photo features the face of a person in the bottom half and the sky in the upper

half, a traditional software may simply display the middle of the photo in a preview. Twitter's AI instead understands that the face is the critical element of the picture and priorities it. [78] Though this might seem like a minor feature, it's quite a big deal. Websites can experience significantly increased usability and conversion rates with this kind of AI.

#43 – AI Curates Music

Ever get stuck listening to the same ten songs over and over again? It happens to a lot of us. Finding new music to listen to isn't always easy. Music streaming giant Spotify has a solution to the problem.

Every Monday, Spotify users receive a unique list of songs in a personalized playlist called Discover Weekly. This playlist is curated by a machine learning AI, which predicts what new songs each individual user might enjoy. The algorithms look at many variables in determining which songs to recommend. For starters, it categorizes the user's taste and matches it with other groups of users with similar preferences in music. Next, it finds songs from those groups that the user may enjoy. It does the same with playlists. If two of the user's favorite songs appear on playlists alongside a third song that the user hasn't heard before, that third song is a likely recommendation.

The AI also reads music websites and analyzes how artists are described in order to match new artists with the user. It can even discover emerging micro-genres like "synthpop" or "chamber pop" and match those genres with users. [79]

#44 – AI Delivers Pizza

The American pizza chain Domino's Pizza began a self-driving pizza delivery pilot program in 2019. Domino's partnered up with autonomous-driving startup Nuro, who supplied self-driving electric vehicles that came equipped with doors locked behind a PIN code. The PIN code is given only to the person who ordered the pizza. Thus, potential pizza thieves can't grab pizzas that don't belong to them. The program started in Houston, though Domino's Pizza intends to do a full roll-out in the coming years. [80]

#45 – AI Takes Pizza Orders

Quite astonishingly, Domino's has placed their AI ordering assistant on virtually every platform imaginable. Through text or speech, users can order pizza using Google Home,

Amazon Alexa, Slack, Twitter, Facebook Messenger, text messages, smartwatches, or even Ford Sync and Samsung Smart TV. [81] The customers do not need to interact with any human when placing the order, as the AI takes the orders autonomously. Users can even save their favorite order and are thus able to make a full order by uttering a mere two or three words.

#46 – AI Scans Pizzas

When customers started to share images on social media of Domino's pizzas that were not looking up to par – with inconsistent or even incorrect toppings – Domino's deployed AI to augment its chefs to more consistently cook excellent pizzas. Using a machine learning AI they call DOM Pizza Checker, Domino's takes a photo of every pizza its chefs cook. The picture is analyzed by the AI, which grades it based on various criteria. For example, DOM Pizza Checker examines if the toppings and cheese are spread evenly and if the selected ingredients are correct for the pizza type.

If the pizza isn't up to standards, the AI asks the chef to remake it. After a satisfactory pizza has been made, the photo is sent to the customer. This way, Domino's ensures that there are no unwanted surprises for the customer when the pizza arrives.

Domino's teamed up with Dragontail Systems in developing the AI. It took two years to launch the first version of the system, which is now self-improving, as it scans pizzas every single day and learns from each and every one of them. [82]

#47 – AI Reinvents Drive-Throughs

Several companies are using AI to reinvent the drive-through experience at restaurants. The giant fast-food chain McDonald's, for example, have started using AI voice-based technologies for ordering food at drive-throughs. [83]

Yet the possible AI applications in drive-throughs extend far beyond simple AI that processes human speech. The startup 5thru uses many AI technologies to reimagine the entire process, through both automation and augmentation. Cameras at the drive-throughs scan the license plate of the customer's vehicle, which an AI then uses to pull up their profile. The AI later provides employees at the restaurant with personalized scripts for upselling to each particular customer, based on their ordering history. Payment for the order is also automated, as customers can register a credit card to their profile, which in turn is connected to their license plate.

5thru claims to be able to reduce the line time at drive-throughs by up to 55% and argues that restaurants will also make more money thanks to the AI-powered upselling recommendations. [84]

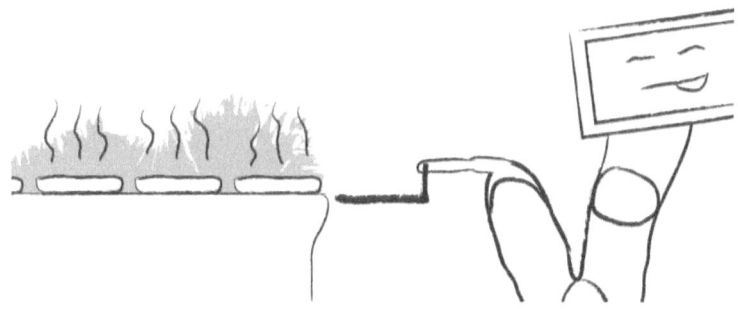

#48 – AI Flips Burgers

Hamburger chain CaliBurger partnered up with Miso Robotics to launch the world's first autonomous robotic kitchen assistant in 2018. Named Flippy, this AI-powered robot is capable of cooking burgers all by itself. While flipping burgers might sound like a rather easy task for a robot that perhaps wouldn't require much intelligence, it's actually a fair bit more complicated than one might think. The robot comes equipped with an image recognition AI, which allows it to analyze burgers and determine when they are finished. [85]

Flippy works side-by-side with humans, ensuring both consistent cooking times and a consistent level of quality. Short-order cooks would often quit after working for just a few weeks, as the kitchen gets incredibly hot. [86] Naturally, the heat doesn't bother Flippy.

A year later, Flippy had learned to cook chicken tenders and tater tots. Flippy also made a new friend: a new robot capable of frying French fries. [87]

#49 – AI Assists Chefs

Samsung's Bot Chef is an AI-powered assistant built to help chefs cook up just about any dish. This versatile robot is basically a robotic arm that is installed in kitchens. It's capable of chopping, whisking, pouring, cleaning, and anything in-between, providing a lot more features than other kitchen-assisting robots. Users can interact with the robot through voice commands and can even program their own skills into the machine learning AI. [88]

#50 – AI Predicts Flight Prices

Hopper is an app for booking flight and hotel tickets. It's in a crowded industry, but Hopper has a terrific competitive advantage. They use machine learning AI to predict the perfect time to book a ticket. The service may recommend a user to book a ticket at a later date when its AI predicts that the price will be lower. According to Hopper themselves, their AI predicts prices with 95% accuracy up to a year in advance. The company also claims that customers can save up to 40% by booking at just the right time. [89]

#51 – AI Distracts Theme Park Visitors

To ensure that visitors to Disney's theme parks are having a good time, Disney implemented an AI that keeps track of theme park visitors. By discovering where tedious queues are forming, Disney can stage a spontaneous parade or offer discounts at less-populated areas to draw visitors away from the queues. [90]

#52 – AI Fights Bullying

For all the great things that the Internet has brought into our lives, it has also brought a lot of bad things along with it. Social pressure is greater than ever before, suicide rates are at an all-time high, and bullying is easier than ever. Many social media giants

have begun to acknowledge the consequences of their platforms. As a response, they have started to adopt technology to fight online bullying. One of them has been more vocal about it than the others: Instagram. Comments made on Instagram are analyzed by a machine learning AI, which discovers if the comment contains a hurtful or otherwise offensive message. The AI stops users who try to post such comments. Through this feature, Instagram hopes that bullies will reflect on their actions. [91]

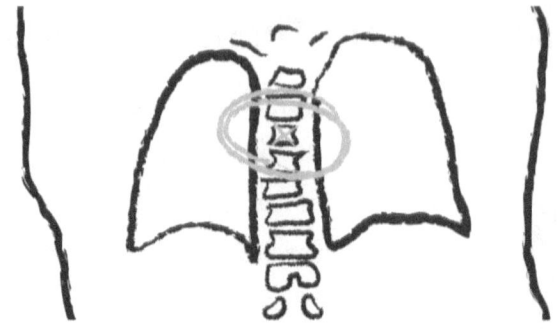

#53 – AI Detects Cancer

Chinese startup Infervision uses machine learning to help hospitals around the world with medical diagnoses. The company assists with 53,000 diagnoses every day for their over 340 partner hospitals worldwide. [92] One such area in which the startup is helping hospitals is cancer detection.

Using image recognition technologies, Infervision's AI can examine an image, detect cancer, and put together a report in less than half a minute — a process that otherwise takes human doctors up to twenty minutes. [93] Furthermore, doctors are often working long hours and are put under large amounts of stress, which can lead to errors in diagnosing. An AI, meanwhile, gets neither tired nor stressed.

In general, AI is both faster and more accurate than humans when it comes to diagnosing, but humans are often still superior when it comes to treating. AI has become somewhat commonplace in the healthcare industry, though rather than replacing workers, it almost exclusively functions as an assistant for diagnosis.

#54 – AI Diagnoses Patients

While some medical solutions opt to specialize in detecting specific illnesses, Babylon Health's AI is capable of identifying many of the most common medical issues. Through a digital healthcare app that can diagnose patients autonomously and provide appointments with human doctors, Babylon is automating medical diagnoses – in general. [94]

A person who is seeking a medical diagnosis will open the Babylon app and answer a series of questions asked by the AI. Based on the answers, Babylon's AI determines the most likely cause of the symptoms. The app is an example of human-machine collaboration, as the artificial intelligence handles the diagnosing of symptoms, while human doctors, with whom the user can also speak with through the app, focus on the treating.

Babylon Health has frequently claimed that its AI can diagnose common diseases as well as human experts, though the startup has received criticism for misdiagnosis in the past. Parts of the UK-based company's software is integrated into Samsung Health.

#55 – AI Diagnoses Parkinson's

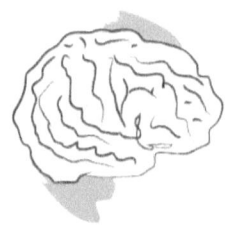

Two Swiss researchers were able to diagnose Parkinson's Disease by allowing machine learning algorithms to analyze smartphone data. By asking people with and without Parkinson's Disease to walk, speak, tap, and perform memory quizzes with their smartphone, the algorithms could learn patterns indicating whether a user had the disease. Parkinson's Disease is often misdiagnosed, an issue that can now be solved by artificial intelligence. [95]

#56 – AI Provides Therapy

If you ever need someone to talk to – whenever and wherever – AI might be a good option. Several AI-powered therapeutic bots have already been released by startups, such as fellow competitors

Woebot [96] and Wysa [97]. These bots provide coaching, therapy, and peace of mind. The treatment they provide is both anonymous and free. Millions of users are already using AI for therapeutic reasons, which shows that while technology may be a significant reason behind mental health issues today, it can also serve as a means to combat these issues.

#57 – AI Recommends Video Content

This is an AI that you may have experienced yourself: Netflix's AI recommendation algorithm. Netflix splits its viewers into more than 2,000 groups, in order to recommend shows and films as accurately as possible. [98] Though many of this US-based company's competitors have similar AI algorithms, Netflix has been tuning their algorithms for over a decade.

Here's a rough explanation of how basic recommendations based on groupings usually works. If a person has seen and enjoyed 20 particular movies, and other people who watched and enjoyed those same 20 movies also enjoyed these other 30 movies, then those movies could be solid suggestions. It sounds simple, but tuning the algorithms is complex.

#58 – AI Detects Fraud

Another AI that has been around for a decade is fraud-detection AI. Paypal began using advanced fraud analytics well over a decade ago. Today, the company can detect fraud in near real-time, thanks to an AI that looks at massive amounts of variables, such  as purchasing activity, the trustworthiness of vendors, and cookies stored in browsers. [99]

#59 – AI Invests In Stocks

In 2017, the Finnish asset manager FIM launched an AI-powered fund, in which a self-learning algorithm selects all 50 stocks included in the fund entirely by itself. By analyzing enormous amounts of varied data, the AI identifies relationships in global markets that humans are unable to detect. [100]

Other AI-powered funds have also started to appear. Coeli Prognosis Machines is a high-performing fund in which financial models are combined with AI to make financial decisions. [101] Another fund – Innolab Capital Index A/S – was developed by a Danish developer to a price of 1.3 million Euros. Targeting professional investors, the AI behind this index fund processes and analyzes vast amounts of data on a daily basis to predict market movements. [102]

Machine learning AI has the potential to run investment funds better than any human. AI can analyze data faster and can find casual relationships that humans cannot. However, it is imperative to note that even fully automated AI funds behave within certain human-defined rules.

#60 – AI Scans Satellite Images

What if you could gain deep insights into anything, anywhere, at any time? Well, a company called Orbital Insight is utilizing that power. The company is applying machine learning AI to satellite imagery to discover astounding insights.

For example, by scanning satellite footage of Tesla's factories, the startup knew that Tesla would miss its financial estimates long before it was publicly revealed. The company knows how much oil there is in the world. They know how many airplanes are at any given airport. They know how many refineries are facing outages. And they know how construction and land are changing in any given area. All of these insights are gained by simply examining changes in satellite imagery. What more, Orbital Insight also tracks supply chains, as they can follow the movement of vehicles and foot traffic.

Helping organizations and governments alike, Orbital Insight uses AI to provide powerful insights on a whole new orbital level.

#61 – AI Mimics Playstyles

Since its inception in the 1950s, video games have used various forms of AI. Every time you play a game that has computer-controlled characters, you could say that you are playing against an AI. However, the AI seen in video games is quite different from the AI that we discuss in this book. Video game AI is almost never built using any form of machine learning. Instead, video game AI has been built with clearly pre-defined behaviors that are determined by human programmers, specifically to provide enjoyment and a consistent challenge to players. The kind of AI described in this book is not usually necessary for video games and is therefore rarely used. Yet in 2014, a developer did release a game with self-learning AI. A car racing game.

Turn 10 Studios and Playground Games created the racing video game Forza Horizon 2 for Microsoft's Xbox One. In racing games, computer-driven opponents often drive similar to each other, making the computer quite predictable, dull, and repetitive. Developers try to add variety by introducing a bit of randomness to the code – allowing the AI to do something unexpected every once in a while. Yet even this behavior is restricted and becomes predictable after extended play sessions.

This is in stark comparison to human players. Every human who plays the game has a unique play style. Compare your friends to each other. One of your friends may brake early when a turn is approaching, while another brakes late. One might ram their competitors, while another friend might opt for cleaner driving. Or what about overtaking: we all have that one friend who likes to stay comfortably in second or third place during the race, only to spurt to first towards the end.

The game's developers wanted to make it feel as though you are playing against humans, and so they implemented a system of racers they call Drivatars. [103] These racers behave much like human players – because they are actually machine learning AI which has trained on humans.

What more, the AI pays extra attention to learning to play like your friends. Even when your friends are offline, and you're playing by yourself versus the computer, your friend might show up. This is an AI that has learned how your friend plays the game and adopted their way of driving. Thus, the game gives you the illusion of playing against friends, even when they're offline.

As machine learning algorithms become even more sophisticated, this type of AI could be applied to more genres of video games.

#62 – AI Enters CRM Systems

CRM stands for Customer Relationship Management. Organizations small and large use CRM systems to interact with both current and potential future customers. As a company that has been very successful in the area of CRM, it may be of little surprise that US-based Salesforce became the first company to build an AI-powered CRM system.

Salesforce created an AI-powered CRM assistant to help its clients be more productive and to speed up decision-making. They named it Einstein, which, for the record, is a rather silly name for an AI. Regardless, the machine learning AI assists users with a wide range of tasks, such as discovering causes of business outcomes, predicting what sales leads will convert, personalizing experiences for customers, and resolving support cases faster.

According to the customer stories that have been highlighted on Salesforce's website, the AI was able to increase lead conversion rates with what they referred to as "top tier leads" by four times. Additionally, the AI increased conversion rates by 9.6%. [104]

Lycka är meningen och syftet med livet, syftet och slutet av människans existens. - Aristoteles

#63 – AI Translates Texts

To be frank, the world's leading translation tool, Google Translate, used to be pretty bad. Around 2016, however, it suddenly started becoming genuinely good. Today, it is considered a quite competent translation tool. This is because Google rebuilt its translation tool with machine learning in 2016 – ten years after the original launch of Google Translate. [105]

Google Translate used to translate each word individually. Today, the AI-powered tool instead translates complete sentences, which allows it to take the semantic meaning of each word within the given context into consideration during the translation process. Following this change, the quality of Google Translate skyrocketed.

Using image recognition AI, Google has also enabled users to perform translations by simply taking a photo of some piece of text. Users can upload a picture that contains text that they wish to translate, after which the AI detects words to translate.

#64 – AI Reduces Pesticides

John Deere is using machine learning AI to reduce the number of pesticides used on crops. [106] They went about doing this by training an AI with images of crops that were labeled as healthy, along with pictures of crops that were labeled as sick. The AI learned to differentiate between the two. Now, it is using this knowledge to inform farmers of when pesticides are necessary to use.

#65 – AI Assists Farmers

John Deere is also using AI to help farmers discover how to better grow their crops in general. They call this AI Farmsight. It's an AI-powered tool that informs its users of

where and when to plant crops. The system collects data from subscribers all around the world, which it then uses to empower farmers with data-driven insights to grow better crops. [107]

#66 – AI Tracks Livestock

Switzerland-based company Viso.ai uses AI to implement vision-based solutions. One of their solutions is an AI that detects the shape and movements of livestock, such as pigs. The company installs AI-powered cameras that both track the animals and provide behavioral analysis. [108] This removes the need to use RFID tagging on animals, which decreases costs, while also granting livestock owners with additional insights.

#67 – AI Picks Apples

Abundant Robotics has created the world's first commercial automatic apple picker. Following major worker shortages among fruit growers, the startup determined to develop a self-driving machine which detects the ripeness of

apples using image recognition AI. The device also picks the apples autonomously through a suction-like piece of equipment. This kind of physical automation is sophisticated and requires expensive machinery, but orchards and farms have already purchased apple pickers from the startup, indicating an apparent demand for the product. [109]

#68 – AI Identifies Suspects

Police forces around the world are using AI to track and identify citizens. This is a controversial area of AI usage, where many judicial and legal landscapes are still in the process of adapting to new AI technologies.

Facial recognition is among the top of the list of technologies with which police forces are experimenting. These AI solutions are used to quickly and automatically identify people who have been suspected of committing a crime. Incidentally, they are also accessible tools for tracking innocent citizens.

At the police department of Washington County in the USA, for instance, officers are using AI for routine errands. An example of such an errand could be when a police officer has received a photo of a suspect, perhaps obtained from a security camera, that they need to have identified. Manually going through their database to find a match could take weeks, perhaps even months. Instead, officers now use an AI to run the photo against their database in search of a match. The AI finds matches at lightning speed, indeed, it identifies the suspect within seconds. [110] The technology is similar to the tech used by Facebook to automatically tag your friends in photos, or the way iPhones can be unlocked by simply looking into the front-facing camera. Except in a police database, you can't opt-out. And you can be wrongfully arrested.

There are many reasons to be concerned about the fact that police are using AI systems. Besides the underlying privacy concerns that come with the technology, discrimination is a major issue. Police AI is routinely being trained on falsified data. Due to a disproportionate targeting of minorities, police data has systematically been skewed to be discriminatory. Police have manipulated data, even downright falsified it, in order to alter crime statistics or to meet arrestment quotas. Police officers have even been found planting drugs on innocents. [111] Now, machine learning algorithms are being trained on this quite terrifying data.

The usage of AI surveillance among police authorities is increasing around the world. While many may have heard of the Chinese government's usage of AI to track its population, the reality is that many more countries are using it – they're simply quieter about it. Law enforcement in at least 52 nations across the globe are using AI technologies for surveillance purposes. [112]

In the United States of America, up to one in four police departments can use facial recognition AI. Many of them use it for routine errands. [110] Its usage varies a lot from region to region. In Orlando, Florida, police tried using facial recognition AI, but ditched the technology, citing technical issues. [113] Another US police department, New Orleans, lied about using it. The police department claimed that they weren't using facial recognition, yet emails revealed that the police weren't telling the truth. [114] Some states or cities in the US have banned AI recognition technologies. The state of California, for example, has banned the usage of facial recognition in body cameras. [115] San Francisco, Oakland, and Somerville have banned the technology altogether. [110]

Dutch police are using AI to catch drivers on their phones [116] – as is Australian police. [117] French police are using AI to track its citizens in an effort to detect unusual movements and formations of crowds. [112] In Hong Kong, where heated protests are occurring at the time of writing, police are also using AI to identify protesters. [118] Facial recognition has also been approved for police usage in countries such as the United Kingdom [119] and Sweden [120], both of which use AI to identify suspects.

There is a wide range of companies offering AI services to police authorities. Chinese giant Huawei is one example of a major player in the surveillance business. In a likely attempt to reach into Europe, Huawei even gifted a surveillance system to France [112], who went on to use it for surveillance purposes. Yet the American e-commerce-born powerhouse Amazon, one of the largest corporations in the world, stands out. As Amazon has slowly turned into a surveillance company, so too has it started to offer a strong selection of surveillance services to the police.

Amazon's controversial facial recognition software, called Rekognition (yes, with a "k"), is being marketed towards police forces. [121] The technology allows for mass surveillance of citizens. In a late 2018 viral open letter, an anonymous Amazon employee urged the company to stop selling facial recognition tech to the police. [122] In the letter, the author explained that 450 Amazon employees had signed a letter to Amazon CEO Jeff Bezos and other top executives, urging them to seize selling surveillance technologies. They argued that the harm caused by their technology could be difficult to undo. "If we want to lead, we need to make a choice between people and profits. We can sell dangerous surveillance systems to police or we can stand up for what's right," the anonymous employee wrote. [122]

Yet Amazon is determined to stay on course. They are even going as far as to write their own laws to present to lawmakers, as if to act like a government instance. However, Amazon does appear determined to pitch responsible legislature: "It's a perfect example of something that has really positive uses, so you don't want to put the brakes on it[,] but, at the same time, there's also potential for abuses of that kind of technology, so you do want regulations," the Amazon CEO Jeff Bezos argued. [123]

According to a 2019 poll, 56% of the US population reportedly trust law enforcement to use facial recognition responsibly. [124] Then again, it's quite likely that the potential negative outcomes haven't been properly communicated to the population at large, so the results may need to be taken with a grain of salt. For perspective, a similar poll was made in the UK, in which people were asked whether they agree with the police using facial recognition technology to identify suspects. Two-thirds of Britons said they were against the usage of such AI by the police. [125] Even British police officers themselves have raised concerns for using AI recognition technologies. They are worried about data bias and amplified prejudices, as officers often found themselves disagreeing with the algorithms. [126]

Police forces around the world are increasingly adopting AI surveillance systems. As the technology can make the police more efficient in catching criminals, many argue that the upsides outweigh the downsides, discrimination or otherwise. Suspects may be identified quicker, though crimes can turn into self-fulfilling prophecies, as overreliance on or misuse of the technology can cause innocents to suffer.

#69 – AI Monitors Hate Speech

Of course, police forces are using AI for more than just image recognition. In the UK, for example, police officers are using AI to monitor hate speech on Twitter. British police discovered an increase in hate speech on social media after the UK announced its intentions to leave the European Union – the Brexit. To combat the increase in hate speech, the police deployed an AI to monitor Twitter posts.

In an effort to more quickly shut down hate speech, machine learning algorithms scour Twitter, searching for anything that could indicate hate speech. The AI flags more than half a million tweets related to Brexit every day, of which less than 1% is classified as hate speech. The machine learning algorithms can supposedly detect hate speech with an accuracy of 85% - 95%.

If the identified hate tweets also have a location tag, a police dashboard will be updated automatically with a map of "hate hotspots." If a hate hotspot grows large, police departments may allocate officers to that area. It's a sort of surveillance system using locations of tweets as a basis. [127]

#70 – AI Recruitment Tool Goes Sexist

One of the biggest threats of near-future AI is discrimination. Out of fear of becoming obsolete, many companies have found themselves rushing to implement AI solutions to stay innovative, cut down on costs, and keep up with the competition. Unfortunately, this rush has led to the creation of discriminatory AI systems, for AI learns from the data that it is fed.

Amazon got to experience this firsthand. They built an AI tool that would assist human resource managers in their hiring processes by automatically reviewing resumes. Regrettably, due to the data pool being dominated by a history of male resumes, the AI learned to favor male candidates.

It might sound easy to teach an AI to simply ignore gender in its determination process, but in reality, it's very complicated. For example, men and women tend to use different words to describe themselves: men are more likely to use verbs such as "executed" and "captured" than women are. The AI learned to favor male vocabulary. [128] Naturally, Amazon shut down their AI when they realized that it had become discriminatory.

#71 – AI Removes Makeup

The popular photo-editing app MakeApp uses machine learning AI to modify photos through all sorts of advanced filters. In 2017, MakeApp released its most controversial filter yet: one that could add or remove makeup to any face in any photo or video. The feature was released to much controversy, as many perceived the app to be misogynistic. The app's creator, meanwhile, argued that the filter was merely released as a fun experiment. [129]

#72 – AI Removes Clothes

Yet MakeApp's makeup-removing filter would be nothing compared to one of the most controversial AI solutions of 2019: Deepnude. Through the usage of machine learning algorithms, Deepnude

allowed users to upload photos of a clothed person, which the AI would then turn nude. The application launched to much controversy, instantly going viral. It was shut down mere hours after release. [130] Given the nature of humans, however, it would be shocking if similar AI services didn't re-appear in the future, next time anonymous.

On a related note, the most common application of deepfakes today is unfortunately fake pornography.

#73 – AI Protects Elephants

You may be familiar with the disturbing and tragic fact that poachers are slaughtering African elephants for their valuable ivory tusks. Thankfully, many groups are working to prevent poachers. Cornell University, for example, is using AI to protect elephants in the Republic of Congo.

Cornell opted to place 50 acoustic sensors in a national park in Congo, to track and count elephants. The sensors recorded all sound in their surroundings, which resulted in massive amounts of data. At first, the researchers planned to use human labor to process the data by manually going through every recording and make a note every time they heard an elephant. However, they quickly realized that it would take years to go through all of the recordings.

Instead, Cornell employed Microsoft to help in creating a machine learning AI to analyze the data. In less than a week, the AI was successfully able to count the number and locations of elephants recorded. Moving forward, Cornell is hoping to extend its AI efforts and run the AI tools directly in African forests. [131]

#74 – AI Fights Animal Extinction

Wildbook is an organization that is fighting animal extinction by blending wildlife research with AI. The organization has adopted a wide range of technologies, one of which is image recognition AI, which Wildbook uses to quickly and accurately perform animal population analyses. [132] Their AI, which was developed together with Microsoft, can automatically analyze individual animals, and provide insights as to where they are and where they have been.

The AI is accurate enough to not only determine that an animal might be a leopard, for instance, but it can also tell whether the leopard the AI is looking at right now is the same leopard that it saw a week ago. It is able to inform researchers how individual animals are moving across regions, while also ensuring that the same animal isn't counted twice.

On top of the AI we just discussed, Wildbook's portfolio of other AI tools is quite vast. For instance, they also use social media to collect information from people who claim to have sighted a particular animal. Their software is open source, so you can check it out yourself.

#75 – AI Writes News

This might surprise you, but many of the news articles that are published today have been written by AI. RADAR (Reporters And Data And Robots) is one organization that is *partially*, in the future perhaps *entirely*, automating news writing. To date, RADAR has created hundreds of thousands of articles, including front-page stories. RADAR has created the world's only automated local news agency, and they supply their stories to hundreds of publications, broadcasters, and websites in the UK. The organization has filed 250,000 articles so far. [133]

RADAR argues that humans *are* still necessary for the process, as expert writers can take a copy crafted by a machine learning AI and add their own level of writing personality to elevate it further. As the AI improves its writing, however, more parts of the news industry could become automated.

For all you know, perhaps this whole book has, in fact, been written by an AI. Maybe this "Jacob Bergdahl" figure is a fictional character whose online presence has been carefully crafted by an AI. Maybe.

#76 – AI Analyzes Phone Convos

No, this story is not about the government perchance listening in on your private phone conversations (are they?), but rather about sales and customer support personnel using machine learning AI to increase their performance.

Cogito is an AI-powered analytics platform that augments such workers by analyzing their phone conversations in real-time. [134] The AI offers real-time guidance by analyzing the tone and energy of the human agent, as well as their level of participation in the conversation.

If the agent is speaking too quickly, or if they need to talk more empathetically or confidently, the AI will suggest such actions in real-time. The AI informs the agents in what regards they are doing well and what they need to improve. For instance, one agent might have an excellent level of energy but needs to speak a bit slower. Of course, the AI doesn't only analyze the agent, but also the client. Cogito tells the agent in real-time if it detects an intention to buy from the client, or if the client is perhaps showing signs of frustration.

#77 – AI Controls Heating

Google's Nest Learning Thermostat is an AI-powered machine learning thermostat that cuts down on your energy bills by automatically adjusting the temperature of your home. It learns heat preferences from its owner by observing their preferred temperatures at various times of the day. Nest also learns what time of day the owner is away from home, during which time it automatically enters power-saving mode. According to Google's product page, Nest saves people an average of 10% to 12% on heating bills and 15% on cooling bills. [135]

#78 – AI Explores Mars

After having developed rovers for planetary exploration for many decades, NASA was ultimately able to build autonomous AI rovers. The self-driving rovers are equipped with terrain detection capabilities, which help them explore and discover significant findings on planets such as Mars. The rovers are also equipped with panoramic cameras, which allow them to take high-resolution images. [136]

#79 – AI Learns Your Face

Face.com has been providing facial recognition solutions to Facebook since 2010, which has given the social media giant the power to identify people in uploaded photos, which in turn has powered many additional features. Though many companies are using such facial recognition technology today, it was unprecedented at the time and gave Facebook a great head start. In 2012, Face.com was acquired by Facebook.com. [137] Many were shocked when Facebook started to automatically tag their friends in photos, though it is now taken for granted.

#80 – AI Writes Ad Scripts

Car manufacturer Lexus wanted an advertisement that was as innovative as their latest vehicle supposedly was, and therefore opted to have an ad script written by an AI. The result was a 60-second advertisement filmed by humans but written by a machine learning AI. [138] You can find the video on YouTube.

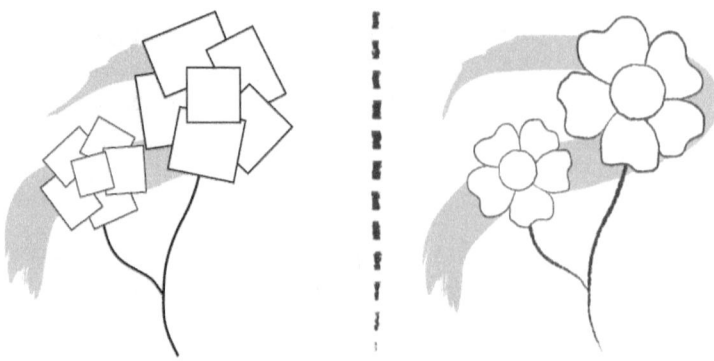

#81 – AI Remakes Video Games

In 2016, a group of researchers invented a ground-breaking machine learning algorithm called SRGAN, later enhanced with ESRGAN. In essence, this AI can take low-resolution images and autonomously make them high-resolution.

This powerful AI was immediately put to use by several communities, including video game fans who wanted to breathe new life into their old favorite games by giving them modern, crisp visuals. For instance, one group of fans recreated the famous video games Resident Evil 2 (from 1998) and Resident Evil 3 (from 1999) using this algorithm. The AI was able to take these 20+-year-old games and upscale their graphics to remarkable results. The people who used the AI to remake the games noted that the AI algorithms didn't produce a *perfect* result, as the AI had an especially hard time upscaling dark areas in video games and also had difficulties upscaling small pieces of texts. Yet, the final result was still impressive. [139] The AI can make it significantly faster and cheaper to perform basic remakes of video games.

#82 – AI Understands Emojis

Instagram taught an AI to understand the semantic meaning of emojis. Though it may seem trivial to be able to interpret emojis, the reality is that about half of every Instagram comment and caption contains emojis. It has effectively created a new language! [140] The path to understanding emojis required quite complex machine learning algorithms, but the result has allowed Instagram to analyze emojis in the same way that it analyzes any written text. This, in turn, let's Instagram target their users with even more accurate ads than before.

#83 – AI Filters Your Face

In 2014, Looksery started releasing AI-powered facial tracking filters, which users could apply to their faces for a good laugh. [141] Acquired by Snapchat a mere year after its launch, the company has now created filters used by hundreds of millions of Snapchat users. Though this technology is often taken for granted today, it was a significant reason for Snapchat's rapid growth in users.

#84 – AI Knows Everything About You

For decades, retailers have been using data analytics to correlate shopping behavior. They use this data to decide how to strategically place items in stores as well as to send targeted advertisements for products that may be particularly effective on each individual customer. Retailers primarily look at shopping history in making these decisions. They are able to determine all sorts of fascinating insights about their customers by just looking at what they're buying. Retailers can tell their customers' gender, whether they live alone, and what time they get off work, by merely looking at what, when, and how customers purchase products.

The massive US retailer Target wanted to take it up a notch and *target* their customers even more accurately with ads. One thing their statisticians tried to figure out was whether or not a customer was pregnant. And sure enough, the company was able to find correlations among pregnant customers. For example, women on the baby registry were buying large quantities of unscented lotion around the beginning of their second trimester. Sometime in the first 20 weeks, they would also purchase supplements like calcium and zinc. They also purchased scent-free soaps and cotton balls.

By using the correlations found in these products, amongst others, Target was able to predict not only whether a customer was pregnant but could also predict the due date. They assigned each shopper a pregnancy prediction score. Should the score be high enough, they would send out targeted advertisements for baby products. [142]

One day, however, Target's algorithm appeared to have made a mistake. An outraged man stormed into a Target store in Minneapolis, USA, demanding to speak with the manager. He showed the manager coupons for baby products, which his high-school daughter had received in the mail. "Are you trying to encourage her to get pregnant?" the father asked. The manager, who naturally had no idea how Target's algorithms worked, looked at the coupons addressed to the teenage daughter. He didn't know why they had been sent to her. He could only apologize.

A few days later, the manager called the man to apologize once more. Yet, much to the manager's surprise, the father was the one apologizing. After the initial confrontation with the manager, the daughter had revealed to her father that she was indeed pregnant.

This infamous and chilling 2012 story demonstrates how even physical retail stores know more about your family than you do. No one could have imagined that a retail chain knew about a family pregnancy before other family members. This story became a wakeup call when it reached the news in 2012. Before this story, few realized what vast knowledge that corporations, even physical shops, had of our private lives.

Since 2012, these algorithms have only become further sophisticated, and as retailers are now sharing information with each other, their knowledge of you has increased considerably. Combine your grocery, pharmaceutical, and clothing shopping behaviors, for instance, and group you together with other customers with similar behaviors, and stores can perhaps even determine your political orientation or your education level.

While many people are starting to become aware of just how much knowledge that Internet companies such as Google and Facebook have about them, many are still blissfully unaware of the insights that retail stores have about their lives. This is amplified by the fact that retailers in recent years have put a lot of effort into hiding their full knowledge of their customers.

Target, the company at the center of this particular story, quickly realized that people got creeped out by what the company knew, and thus started to disguise their insights more cleverly. They got more subtle in how they went about sending advertisements by making their coupons appear more random. They began to provide ads for products they knew pregnant women would *never* buy – like an ad for a lawnmower – next to advertisements for products they *would* buy, such as diapers, in order to make it look like the products were selected by chance. [142]

#85 – AI Recommends Products

Obviously, it's not just physical retailers who are recommending products based on shopping behaviors. Online stores do it as well, and they are much better at it, as they have a lot more data to feed their AI.

Online stores recommend products not only based on the shopper's past purchasing history but may also look at variables attained from social media and Google analytics, in order to determine what the customer's interests may be. They can look at location data and even what web browser the users are using to puzzle together correlations as to what the shopper might be prone to buy.

Any online store that doesn't use machine learning to recommend products is likely doing something wrong. Conversion rates can be increased dramatically by offering the right product to the right customer, and basic product recommendations don't even need to be particularly complicated. Amazon, for example, is one of the masters of product recommendations. They even offer their help to other stores through Amazon Web Services. [143]

#86 – AI Finds New Store Locations

Some companies are using AI to determine where to open new stores. Among them is the coffee giant Starbucks. Taking factors such as traffic patterns, population density, average population age, crime rate, neighborhood income, various demographics, and proximity to other Starbucks stores into consideration, the machine learning AI provides advice for prime locations to open new stores. [144]

#87 – AI Blocks Spam Email

A machine learning algorithm that you (perhaps unknowingly) use every day is a spam filter. Many spam filters use machine learning to determine which  emails are spam and which aren't. Included among them is Google's Gmail, which has perhaps the best spam filter in the world. Their spam filter learns to detect whether or not a particular email is considered spam based on emails previously reported by users. The AI is so successful that less than 0.1% of email in an average Gmail inbox is spam, according to Google themselves. [145]

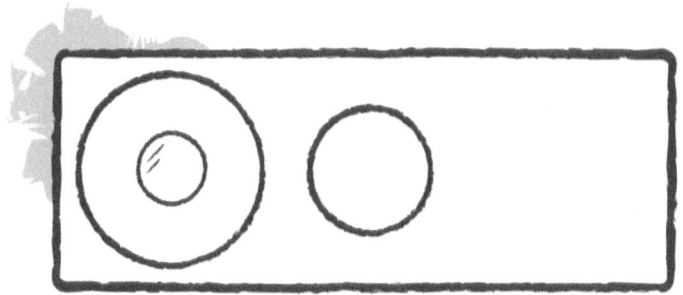

#88 – AI Tracks Motion

Microsoft's Kinect devices are a series of motion sensing hardware that are used to detect how human bodies are moving. Kinect was initially released in 2010 in the form of a peripheral for the video game console Xbox 360, built with motion-tracking tech, as a direct response to compete with Nintendo's Wii gaming console.

Kinect was initially made for video games where the player moves their body, such as dancing games. In more recent years, however, it has been iterated for enterprise software, powered by Microsoft's AI platform Azure.

Azure Kinect AI started shipping in mid-2019, almost ten years after the original release for Kinect. Today, Kinect is being used across a wide range of industries. In health services, for instance, Kinect is used to monitor athletic performance and rehabilitate patients faster based on feedback from the body tracking system. Kinect is also designing better shopping experiences in retail, dispatching items more efficiently in logistics, and automating tasks in robotics. [146]

#89 – AI Powers Mass Surveillance

AI is being used as a method for mass surveillance across the world. At least 75 nations are using AI for surveillance purposes, most of which are liberal democracies. [112] However, semi-autocratic countries are more likely to abuse AI surveillance technologies. Military spending is closely correlated to AI surveillance spending, as 40 of the top 50 military spending countries in the world also use AI surveillance tools.

One nation stands out above the rest. One nation has established a considerable network of mass surveillance that scores citizens based on their actions.

Ever since the establishment of the People's Republic of China in 1949 and Mao's rise to power, China has been conducting mass surveillance. Though it's hard to pinpoint when AI came into the picture, machine learning AI has now become a central method for China to monitor its citizens. Today, the country employs a nation-wide Black Mirror-like social credit system in which all citizens are given a score that shifts depending on their actions. First announced in 2014, the system has grown to include larger and larger amounts of citizens. In 2019, China finally forced all of its citizens to join the system by requiring them each to scan their faces before being able to access the internet. Though China is not alone in employing mass surveillance, it is the only country in the world where citizens are required to scan their faces before being able to sign up for mobile and internet services. [147]

The exact method for how the social credit score moves up and down is kept secret. However, examples of actions that have caused infractions include bad driving, smoking in prohibited zones, traveling without a ticket, buying a lot of video games, posting fake news online, [148] blocking the sidewalk, jaywalking, and loitering. [147] These activities are tracked through cameras placed around cities. The AI-powered cameras not only monitor people by their faces but are also intelligent enough to have learned to identify people by their walking styles. Thus, individuals cannot hide from the AI even if they're covering their faces. [147]

Citizens with low credit scores receive various punishments. For instance, they may lose the privilege to travel by plane or train. Already by the end of 2018, citizens had been denied the buying of plane tickets 17.5 million times, and the buying of train tickets 5.5 million times. [149]

Low credit score holders may also have their internet speed reduced. Additionally, they may find themselves and even their kids blocked from being able to enroll in the best schools. Unsurprisingly, they will also be unable to get good jobs at government-controlled companies, and they may also be unable to stay at decent hotels. Pet owners may even have their pets be taken away from them if their social credit score becomes low enough. [148]

The Chinese government's usage of AI is one of the largest-scale AI operations in the world. Much of that AI technology is provided by Chinese startup SenseTime: the highest evaluated AI startup in the world. SenseTime has been appointed as the leader of China's group for setting standards in facial recognition. [150]

On top of surveillance through cameras, China also performs extensive internet censorship, often nicknamed The Great Firewall of China.

This internet censorship is in large parts fueled by AI technologies. By law, the government of China can remove any content from any online website or app and block any platform that doesn't allow for this censorship. This is one of the main reasons why most western companies, such as Google and Facebook, are banned in China. Today, Facebook and Google do not provide the Chinese government with the means to censor content online. However, for media companies that do allow for government surveillance, the government has deployed AI bots that scan content and either autonomously deletes content that it dislikes or notifies human personnel for decision-making.

AI is enabling China to execute mass surveillance and mass censorship at levels that have never been possible in the past.

#90 – AI Monitors Bus Drivers

On top of the general nation-wide mass surveillance deployed in China at large, individual jobs are also monitored in different ways. Bus drivers are one example.

Some bus drivers in Shenzen, China, are being monitored by AI. If the bus driver is chatting with other passengers, is looking at their phone, is about to fall asleep, or is performing other unsafe behavior while driving, the AI will display an alarm and notify both the driver and their managers. The system uses infrared technology, which means that the cameras watching the bus drivers can see their eyes even if the drivers are using sunglasses. [151]

The purpose of the cameras is allegedly to reduce accidents, not only by enforcing behavior from bus drivers but also by detecting which bus routes cause more fatigue than others. Whether this tracking is connected to China's overarching social credit score system is unknown, but alerts *are* stored in a central system. As of early 2019, the system had been installed in over 700 buses in Shenzen.

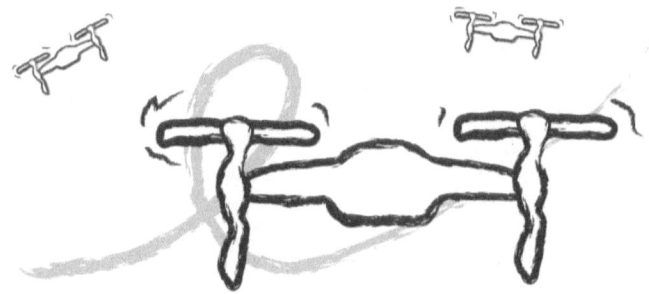

#91 – AI Attacks Drones

A drone is a popular type of an unmanned aerial vehicle. Though drones have many benign applications, they are not always fun and games.

At airports, drones can cause delays that cost millions of dollars to airports and airlines in lost revenue, as aircraft aren't allowed to take off if an unidentified drone is inside of an airport's restricted airspace. Data centers and corporate campuses, meanwhile, are a common target for espionage through drones. Oil refineries are in danger of hostile drones, given their vulnerable infrastructure. In prisons, drones are sometimes used to delivers goods such as money or weapons. In stadiums, drones can deliver weaponized chemicals or explosives. [152]

In 2018, the president of Venezuela was targeted by numerous bomb-equipped drones during a speech. The drones were unsuccessful in harming the president, though several members of the national guard were hurt during the attack. [153]

These are a few examples of how drones can cause substantial damage. As a countermeasure, US-based Fortem Technologies is offering AI-powered drones to combat hostile drones – a sort of drone on drone scenario. For the rest of this story, let us refer to Fortem's drones as "good drones" and unidentified scary drones as "bad drones."

The good drones can autonomously detect, pursue, and capture bad drones using nets, and they can do so from a safe distance.

Fortem Technologies has built an AI platform that fuses radars with other sensors to automatically monitor any environment in a 3D environment. This platform is used together with the good drones to identify bad drones. Naturally, the platform doesn't only detect drones, but also airplanes, wildlife, and other vehicles. It also generates various insights to be used for further analytics.

Additionally, Fortem Technologies offer portable airspace security solutions. These wireless systems track and detect all objects around them, including both people and vehicles. They allow for safe and quick identification of whether the identified objects are friendly or rogue. The system is easily mounted on a tripod, a vehicle, or a hitch. [152]

Backed by Boeing, Fortem Technologies have created a whole new market, in which it ironically protects people from a very similar product. As drones are a relatively inexpensive and accessible way of creating powerful tools for espionage or harm, Fortem's mission is as urgent as it is crucial.

#92 – AI Helps Real Estate Investors

Skyline AI is offering real estate investment machine learning. Using algorithms that study owner and asset behavior, the company identifies which physical areas are about to take off. The AI is fed data from more than 200 data sources. It finds value creation opportunities in real estate, as well as distress signals and mismanagement anomalies. [154]

#93 – AI Powers Vending Machines

Japan is absolutely packed with vending machines! When I was living in Japan, I learned that there is about one vending machine for every 23 people. Given the popularity of these machines, it was only a matter of time before they would become empowered by AI. Indeed, some of Coca-Cola's vending machines in Japan use AI to promote specific drinks for specific areas autonomously. The AI can even offer unique discounts. The machines may change their current promotions depending on their positioning and on the data that they collect from their customers. [155]

#94 – AI Detects Image Manipulation

Realizing the ethical implications of their actions, Adobe, the inventor of famous photo-editing software Photoshop, built an AI for detecting whether an image had been manipulated.

In a 2019 post made on the company's own blog, the Adobe communication team argued that fake content is becoming an increasingly pressing issue, and that trust has become important as image editing has become abundant. Thus, the company began to trial various innovative solutions to increase confidence in digital media. They ultimately landed on AI. [156]

The company trained an AI on thousands of pictures, in order to teach it to detect whether an image had been digitally altered. Eventually, the AI became better than humans at spotting fakes, even reaching 99% accuracy in a series of experiments.

It cannot be understated how crucial AI such as this one is! It will help to democratize image forensics and combat falsified news. Even if it *will* make Instagram influencers sweat a little.

#95 – AI Measures Natural Disasters

One Concern has built an AI platform that allows governments and organizations to mitigate, prepare, respond, and recover from natural disasters. [157] According to the company itself, the number of natural disasters in the world has quadrupled since the year 1970. In the last 40 years alone, natural disasters have claimed the lives of 3.3 million people and caused damage to a cost of 2.3 trillion US dollars. [158]

The company is building long-term protection against natural disasters, using machine learning to help create more resilient infrastructure and communities. Their AI platform offers a dashboard that contains both maps and visualized data, showing near real-time impacts of disasters to cities and neighborhoods. The tangible insights provided by the machine learning AI help humans make better decisions. [157]

With climate change increasing in urgency for every day that passes, these kind of AI solutions appear more and more critical.

#96 – AI Assists Customers

Millie is a large screen – a little taller than the size of an adult human – packed with a context-aware AI avatar. Think of it as an AI assistant, but with a large screen and camera, and made specifically for business-to-consumer assistance.

Millie has been developed by Twentybn, who describes it as the world's first AI avatar. The large screen-with-AI can be used as a personal trainer, a brand promoter, or a style advisor, to name a few of its uses. As an assistant powered by cameras and facial recognition tools, Millie is able to recognize over 1,000 dynamic gestures in real-time and comprehend over 120 languages. [159]

For example, if a customer needs help in-store with figuring out which pants to buy or if they want to know whether those shoes that they saw last week are still in stock, they can turn to Millie. Similarly, if a customer enters a gigantic shopping mall in search of a particular store that they cannot find, they can ask the AI to display directions. Arguably, non-AI-powered screens can accomplish all of the above features as well, but hey, they can't compliment you on your smile, can they?

#97 – AI Creates Valuable Art

Would you believe me if I told you that a painting made mostly by an AI was sold at an auction for 432,000 USD? [160] Who's kidding who, you're not surprised by that fact, you've heard stories of downright blank paintings being sold for millions, haven't you? No price is too high for art connoisseurs.

Anyway, the painting depicted fictional character Edmond de Belamy and was indeed sold for a remarkably large amount of money in 2018, creating a new chapter in AI art. However, while many news publications referred to the art as being made entirely by an AI, the truth is that there were obviously some humans involved in the process. Humans created the base upon which the AI could run its algorithms. More specifically, three French students who collaborate under the art collective called Obvious made the painting by deploying a famous algorithm called GAN (that's short for *generative adversarial network*, it was invented by Ian Goodfellow and colleagues in 2014). [161] It is not clear just how much of this work of art was made autonomously by the algorithm and how large of a hand that humans had in the process.

#98 – AI Judges Gymnasts

During the World Artistic Gymnastics Championships of 2019, judges used AI to help with the scoring. Combining advanced AI algorithms with 3D sensors, Fujitsu created the AI to help with the scoring of the participants. The gymnasts had their bodies scanned before the event started. Later, during the competition, 30-something small boxes were placed around the gymnasium. These sensor-equipped boxes captured the gymnasts every move, analyzing skeletal positions and speeds. [162]

For this competition, AI was being used to assist human judges, though many speculate that AI could soon replace judges altogether, as the AI is able to score competitors more objectively. It is also worth mentioning that Fujitsu believes that, moving forward, the AI can be used for more applications than just judging competitions, such as helping athletes with training. [163]

Fujitsu now has its eyes on the Tokyo 2020 Olympic Games gymnastics competitions. [164] Whether the AI will be allowed to judge at the Olympic Games remains to be seen.

#99 – AI Tracks Checkout Theft

Retail giant Walmart is using image recognition AI to monitor checkouts and prevent theft in more than 1,000 of their stores. AI-powered cameras survey checkout lines, monitoring and analyzing activities at both self-checkout stations and staffed checkout counters. [165] If the AI detects a potential issue, it notifies members of the staff to intervene. Examples of such problems could be a customer forgetting an item at the bottom of their cart, cashiers who accidentally didn't scan an item due to being tired or distracted, or, naturally, theft.

Retailers in the USA lose billions of dollars every year due to theft, scanning errors, and fraud. This AI investment by Walmart is an effort to reduce the amount of money lost to such causes.

This is just one of many innovative solutions developed at Walmart. The company has set up a large unit dedicated to innovating the retail experience with modern technologies. [166] With increased competition from Amazon, innovative solutions are a necessary step for the survival of retailers.

```
vcAnuU (vnpsc) {
    ('b)ign<cc;
    ('.b-cc(gUUb2.<-g;
}
```

#100 – AI Empowers Developers

Enter the last story of this book. AI is empowering software developers and designers in a multitude of ways. I am a developer myself. I have built plenty of websites and apps using a wide range of programming languages and tools. Throughout my years of creating digital solutions, I have discovered many ways in which AI could be used to make my work easier. Luckily, I am not the only one to realize that potential. Let's look at a few of the AI solutions that are available for software developers already.

Software developers write their code into some form of application. These tools are called developer environments. A developer environment can be very simple, though when writing large amounts of intertwined code, developers prefer to use sophisticated software. One of the kings of developer environments is Visual Studio, a Microsoft-owned piece of software that supports many common programming languages, including C, C++, and C#. Modern developer environments often come with a built-in auto-complete-like tool that allows developers to write code faster and cleaner.

These tools can convert short-hands into full code snippets and refactor code to make it more reusable. Powering these tools with AI was a natural next step.

Since 2018, Microsoft has employed IntelliCode to recommend method calls (a method is a programming module that executes some lines of code) based on the context in which the method call is being used. [167] In late 2019, Microsoft announced that they're further enhancing developer productivity by supplying AI-powered refactoring and suggesting, as well as automatic code completion. [168] Of course, Microsoft isn't the only one who is developing such AI productivity tools. Kite is doing the same thing for Python environments [169] and Codota as well for Java environments [170], to name two other examples. Python and Java are both programming languages.

Microsoft uses AI to empower developers in other ways as well. Through its initiative called AI Lab, Microsoft has created a software called Sketch2Code. [171] This powerful tool uses AI algorithms to convert sketches into code. The user can simply draw a sketch of a page of a website that they want to create, upload the sketch to the software, and the AI will generate so-called HTML code (HTML is a form of markup language used to give structure to websites). In its present form, the code generated by the AI is not good enough to be deployed without having a human developer step in and make quite severe adjustments to the code. However, it is useful for generating a basic structure to use as a starting point when crafting a website. In the future, as the AI becomes more sophisticated, it could automate the coding of basic website designs altogether. Microsoft has competition in this area as well, as rivaling tools such as Uizard's Code2Pix are providing similar features. [172]

Another AI provided by Microsoft that I must mention (Microsoft sure does create a lot of tools for developers), is PROSE. That's a strangely put together acronym that one way or another stands for Microsoft **Pro**gram **S**ynthesis using **E**xamples SDK, which is an unusually technical name considering that developers generally prefer to give things names that sound like Pokémons. Anyway, it's an AI for developers that is capable of generating code by being fed examples of input and output. [173] PROSE exists in a new domain in AI called Programming by Examples (PBE). Without getting too technical, PBE is a means of generating code by simply providing example inputs and outputs.

Let me break this down with a basic example. If a developer wants to reformat the way they display the names of their users from "FirstName LastName" to "LastName, FirstName," they can simply tell the tool that they want "Jacob Bergdahl" to come out as "Bergdahl, Jacob." The solution will then generate the code necessary to make that transformation. PBE can supposedly increase developer productivity by 10 – 100 times. [174]

DeepCode is another solution that empowers developers through AI. After a developer has finished writing some piece of code, the code normally has to go through a code review, performed by someone other than the author of the code. Code reviewing is a sort of quality assurance process that ensures that the code is of good quality. Code reviewing is time-consuming. Naturally, as is often the case when a rules-based process is time-consuming, machine learning engineers are seeking to automate the process. DeepCode, to name an example, specializes in discovering security vulnerabilities in code, claiming to have a precision rate of 90%. DeepCode's AI bots not only review the code, but they also provide suggestions for how to fix security flaws. [175]

AI is also assisting developers to more easily make their websites accessible to people with disabilities. AccessiBe, for example, uses machine learning to scan websites and add customized accessibility tools. Developers need only add one line of code. [176] In my own trial of their AI, I observed that the machine learning they use is quite primitive so far, but has a lot of potential. In many regards, online accessibility can be almost entirely automated, and this AI is a part of the process.

Finally, I feel obliged to mention that developer environments for creating AI solutions in and of itself are becoming more and more straightforward. Such platforms include Google's AutoML, Baidu's EasyDL, and Microsoft's Azure Machine Learning.

As software development is becoming increasingly AI-powered, so too is the learning curve for programming getting smaller and smaller. Hopefully, this will encourage more people to try software development.

PART THREE

ANALYZING REAL AI

Creating Value With AI

Whew! You just read through a hundred stories of real-world implementations of artificial intelligence! But we're not done yet. Without getting technical, I would like to show you how to understand artificial intelligence better. I will present a framework that you can use to either analyze the value of AI solutions or to create value with AI yourself.

To be clear: no one adopts AI for the sake of adopting AI. Decision-makers always look for solutions to problems. Whether or not that solution happens to be AI is generally irrelevant. This is why frameworks such as the one that I am about to present are so important. They show how AI could solve a problem, and they generate a strategy in the process.

The framework is non-technical and business-facing. It's a tried-and-tested, easy-to-apply framework that I have used extensively in both my academic and corporate life. I'll go over the origin of the framework first, after which I will guide you through its concepts and usages.

After explaining the framework, I will explore some trends based on the stories that we covered previously. Did you notice that Domino's Pizza has applied three completely different AI strategies to their business? Or that AI is increasingly being used as an enabler for product development?

Let's go a little deeper into the strategies behind AI.

A Framework For Creating Value With AI

I would like to share a fantastic framework that you can use to analyze and apply AI. This framework is actually one of the reasons why I got my first job as an IT consultant. It's the premise of one of my most-read articles ever, and also the premise of my most popular webinar ever. I have used this framework in seminars and lectures. Executives and managers have responded very positively to the framework, saying that it made them feel a lot more comfortable with artificial intelligence. Even the US Army AI Task Force told me that the framework helped them to convey a very complicated topic to executives. It's a tried and tested, easy-to-apply framework.

The origin of the framework

I first discovered this framework when I was writing my master's thesis on the subject of creating business value with AI in the context of enterprise resource planning (ERP[3]) systems (exciting, right?). The framework was made by the consultancy firm Accenture in 2016 [177], though I verified it with extensive amounts of research from MIT and evolved it a bit myself. The academic world was pleased with my rework of the framework and rewarded me with the highest distinction available for the thesis, which propelled me into a

---

[3] An ERP system is an off-the-shelf streamlined IT system that can span an entire organization.

corporate career. Look, what I'm trying to say is: it's a pretty useful framework.

My version of the framework looks like this:

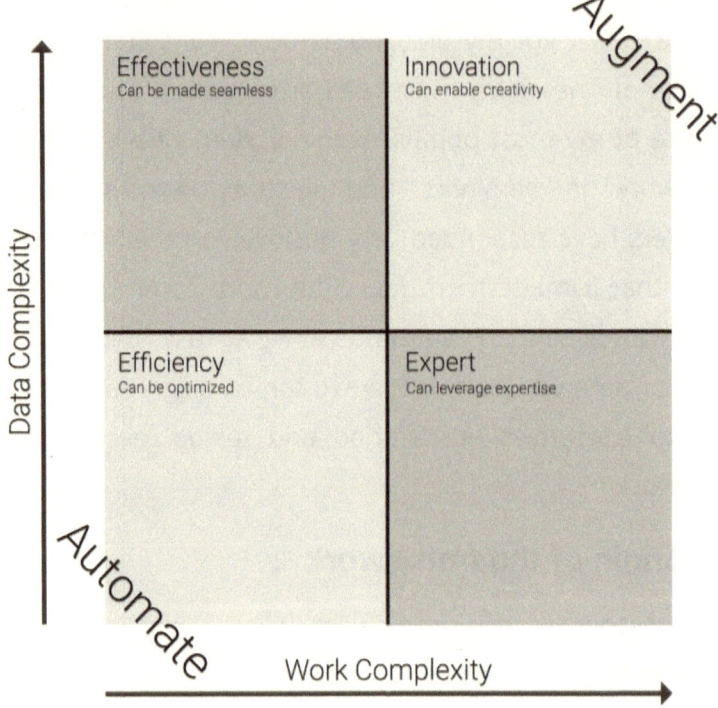

We'll get back to this image in a bit, but first, let me walk you through the framework step by step.

Presenting the framework

The framework is used for analyzing what value artificial intelligence can create. It is a first step that helps decision-makers determine which AI strategy to adopt for any given process (or activity). The central thesis for the framework is that **AI can create value for any given process through either automation or augmentation**.

It's crucial to understand this central thesis, so let's break it down into pieces and examine each critical term in it.

First: process. In this context, a process can be any activity. Living life is an example of a process that we are all performing, though it is obviously too high-level of a process to apply to anything useful. Processes can be broken down into subprocesses, and they are unique to each individual or each organization. Working is a process. A subprocess of working might be to work in human resources, in which recruiting is another subprocess, which in turn has subprocesses such as creating job advertisements, reading resumes, and interviewing. We'll be discussing processes more extensively in a little bit, but the key thing to note is that a process could be practically any activity.

Next: automation and augmentation. Automation means to remove a person from a process, while augmentation means to empower a person in a process. These are, essentially, the two actions that AI is capable of performing.

Right. So, AI can improve any process through either automation or augmentation. But how do decision-makers know which one to apply? When should one choose to automate, and when to augment?

Data complexity and work complexity

According to the framework, the decision to automate or augment is determined by two variables: data complexity and work complexity.

These two variables can be represented in a matrix as such:

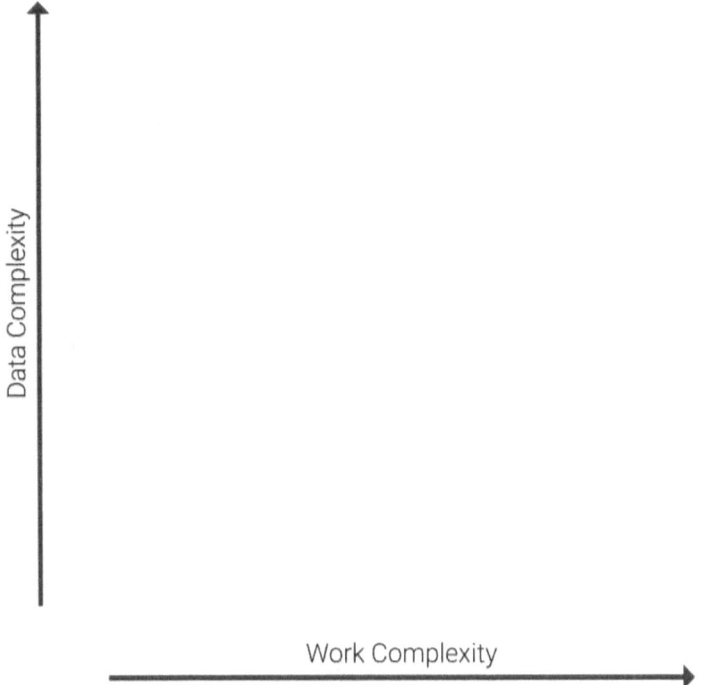

By examining the data and work complexity of any given process, one can determine whether automation or augmentation is the appropriate option.

Data complexity is the easiest of the two variables to explain. Data with low complexity is typically structured and simple: these processes usually consist of simple pieces of text or numbers. This data is easy to interpret for a computer. High data complexity, on the other hand, is often unstructured and up for interpretation. Images, videos, music, and voices are examples of complex data. For instance, it can be up to subjective interpretation to determine whether a person looks tired, annoyed, upset, or are just resting their face. Even we humans may have to ask the question: "Are you upset?"

Work complexity is all about determining whether or not a process has clearly defined rules and routines. If a process does have rules and routines, it becomes predictable and therefore has a low work complexity. If the process is generally unpredictable and ad hoc, however, it requires judgmental decision-making skills, which would result in high work complexity. Computers have historically been excellent at predicting, but not so good at judging.

Processes consist of tasks. Most tasks have a predictive step and a judging step. For example, driving a car is a process, and making a left turn is a task. The predictive step is turning on your blinker and getting into position to determine whether or not it is safe to turn left. The judging step is the act of actually determining whether or not it is indeed safe to make the turn. Computers are much better than humans at the first step, but generally not so good at the latter step, especially in more complex processes. Tasks that contain a difficult judging step are usually of high work complexity.

It is important to note that time is not really a factor in determining work complexity. A process with low work complexity could take a year to execute, while a process with high work complexity could take just ten seconds. What matters is whether or not the process can be handled without, or at least with very little, human input.

As you might imagine, if a process has low data and work complexity, it is suited to be automated. If a process has high data and work complexity, it is suited to be augmented. We can update the matrix with these two actions:

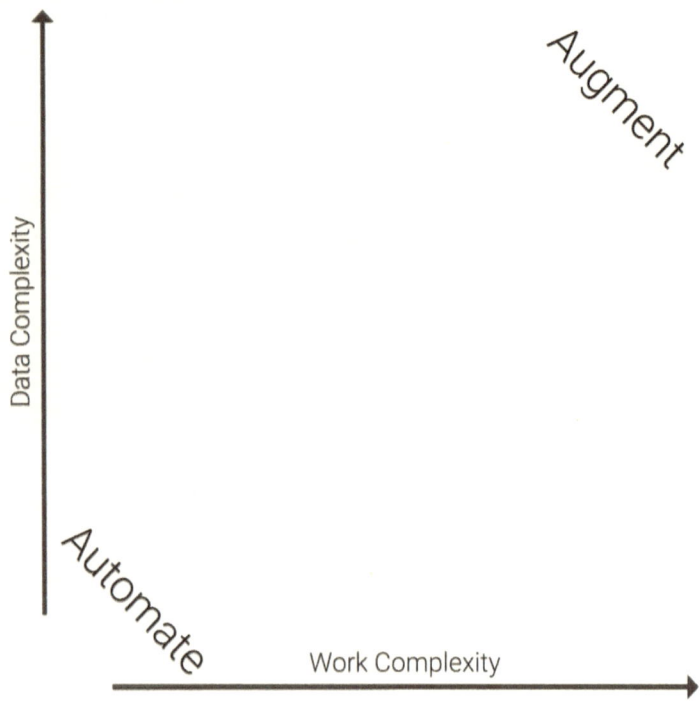

The further to the bottom-left a process is placed, the more likely it is that it should be automated, and vice versa.

The framework doesn't just stop there, however. Depending on where a process is placed in the matrix, we can empower it using one of four strategies: efficiency, effectiveness, expert, and innovation.

These are the four possible AI strategies. Every AI application that exists can be mapped into one of these four strategies. The choice of strategy is determined by the data and work complexity of the process.

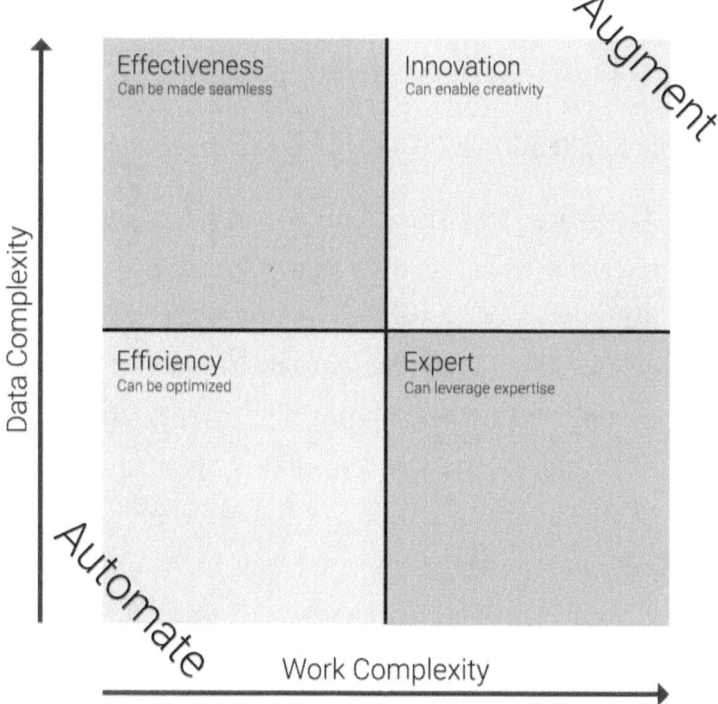

Let us go through each of these four strategies in detail.

The Efficiency Strategy

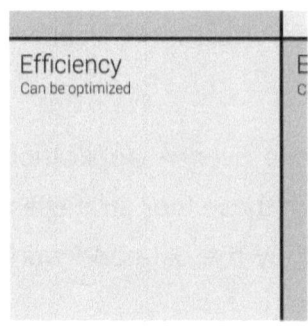

If a process has low data and work complexity, it can most certainly be automated. This is the simplest strategy, and the one that most companies are trying to adopt today. It is a clear first step for any company wishing to start their AI journey. Efficiency is all about optimizing activities, generally with the intent of reducing costs. Processes that are placed here have very clearly defined rules and routines.

Example applications:

- Automated fraud detection. AI can automatically make decisions based on data anomalies (#58).
- Automatic purchasing of standard products from wholesalers. AI can look at any number of variables, including stock level, upcoming events, the season, and other factors, to determine how many packages of milk or eggs it should order this month.
- Automated recruitment. By having AI bots analyze LinkedIn profiles, organizations can automatically invite appropriate candidates to interview's through personalized invitations.
- Autonomous content creation. Any form of text can be written autonomously, be it a content description (#41), a summary of an article, or even a full news article (#75).

- Automatic package delivery. Be it through trucks, drones, or other methods; AI can drive many types of vehicles autonomously. Warehouses can also be automated, with robots packaging products to send to consumers or stores (#40).
- Autonomous managing of stores. AI can manage stores without the need for any human personnel (#22).
- Automatic financial investments. AI is capable of autonomously making financial investments, such as picking stocks for an investment fund (#59).

Some companies take efficiency to entirely new levels, automating most, if not all, of their business. Richard Liu, CEO of Chinese e-commerce giant JD.com, hopes that his company will one day become 100% automated by robots, for instance. [39] It will take a very long time before that happens, though, if it ever does. Most processes today simply cannot be automated. AI isn't agile or general enough to perform most of the tasks that humans are capable of; neither does it have the cognitive ability for many of the unexpected challenges that decision-makers are faced with today. On top of all this, AI solutions are sometimes more expensive to employ than humans. Thus, automation is sometimes pointless for corporations who are just trying to cut down on costs.

One final thing to note about this strategy is that organizations usually do not automate jobs with the intent to fire employees, but rather to re-allocate employees into more cognitively challenging tasks.

The Effectiveness Strategy

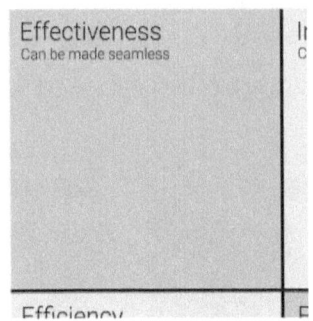

If the work contained in a process is simple, but its data is complicated, the suitable choice of strategy tends to be the effectiveness strategy. This approach revolves around the communication and coordination of workers, with AI serving as the role of an assistant. Processes placed in this strategy are used to make humans more effective, by either eliminating or simplifying the act of scheduling, communicating, and/or monitoring. NLP is a common AI technology used to empower processes with this strategy.

Example applications:

- Autonomous scheduling of meetings. AI can schedule meetings for humans based on their availability or preference. Meeting notes can also be generated automatically, by first having an AI autonomously transcribe the meeting, and then having it create a summary.
- Streamlined communication. Some meetings can be avoided altogether, by having AI observe human workers, and explain their work to relevant team members, at a level they understand. For instance, experts and novices in the same field of work may have different levels of understanding of the same topic, a gap that AI can help to bridge.

- Automatic determining of which consultant is a good fit for a particular task. In large consultancy firms, it can be challenging to know just which consultant to send for any given project. AI can help by selecting consultants based on their experience level, preferences, skills, goals, availability, and cost.
- Automated customer support. A classic example that has been widely implemented already. AI is often the first line of support today, as it can answer and help with most of the common support errands that organizations receive. If a support task is too complicated for the AI to process, a human agent can join the support errand.
- Placing orders through a digital assistant. Rather than calling a human to order a pizza, a person can simply communicate with a virtual assistant (such as Google Assistant, Facebook Messenger, or Amazon Alexa). The assistant informs the chefs of the order – no middle-management required (#45). This can be applied to any business.

Few companies truly utilize the full potential of the effectiveness strategy. Companies that work extensively with this strategy will find their employees spending more time conducting meaningful work, and less time on communicating and coordinating predictive work-related errands. This is not about preventing humans from communicating with each other, but merely to streamline the boring, slow, or difficult parts of communication.

The Expert Strategy

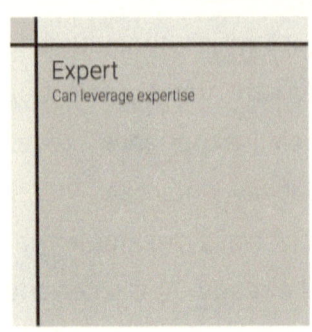

Expert
Can leverage expertise

When the data is simple, but the work complexity is high, AI can be applied to leverage expertise. Processes placed in this strategy rely heavily on decision-making judgmental abilities. Unlike the left-most AI strategies (efficiency and effectiveness), humans make all decisions in these processes. AI enters a supportive role, opting to offer advice and insights. In this strategy, humans are always in control. Expert systems deal with anything from large amounts of money to human lives, meaning humans must always be responsible for the consequences of decisions made.

While the left-most strategies mostly revolve around reducing costs, the expert strategy revolves around increasing revenue, boosting performance, and providing new competitive advantages.

Example applications:

- Medical diagnosis. By entering the symptoms of a patient, an AI-powered system can suggest potential causes of the patient's illness and recommend appropriate treatments (#54). In the future, medical diagnosis could be automated altogether. However, in today's world, AI in healthcare is typically used as an assistant to human experts.

- Selective purchasing from wholesalers. Niche shops that sell exclusive items may use expert AI systems to find products that will appeal to their customer segment.
- Financial investment guidance. Financial AI can assist investors and economists in making decisions regarding investments. Progressive firms have automated some financial activities already; however, given the amount of money that is often at stake, AI in finance is typically augmentative in nature.
- Support consumer decision-making. For instance, there are online travel fare aggregators that use AI to predict how the price of flight tickets will change over time, advising its users to either buy now or wait for a lower price (#50).
- Assist in inventing new products. Specialists in product development can be assisted by an expert AI in the process of developing new products. On page 164, I present a trend analysis on this very topic.

As routine-based jobs become automated through the efficiency and effectiveness strategies, companies will need to help employees gain the necessary skills to execute more complex tasks. Organizations can accomplish this by investing in the expert strategy. By making complicated jobs more straightforward, humans can more easily make the transition from performing rules-based duties to executing unpredictable jobs instead.

The Innovation Strategy

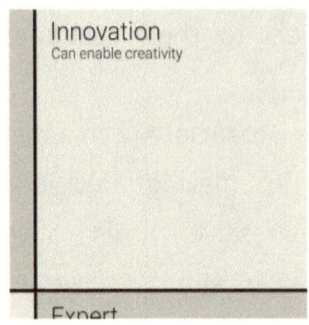

Innovation
Can enable creativity

Expert

Processes that have both high data and high work complexity can be used to enable creative work through the innovation strategy. Here, humans are in complete creative control. AI is deployed to be an assistant that identifies suggestions and alternatives. As you might imagine, this strategy is particularly interesting for jobs that are creative or innovative in nature, such as designing, researching, writing, cooking, composing music, drawing, filming, and so on. Yet most jobs do, in fact, contain some amount of creativity. For instance, every time a person writes a text message to a friend or has a conversation by the coffee machine, they are actually performing a creative task.

Example applications:

- Augment interviewers. While interviewing a job candidate, an HR employee can be augmented with an AI that suggests follow-up questions to ask the candidate in real-time, based on the conversation the two are having.
- Augment music composers. An AI might recommend a music composer to add an appropriate bass to a song depending on the other instruments that the composer has added. An AI could also create samples for humans to mix (#34).

- Augment writers. When a user writes some piece of text in a Microsoft Word document, the software may recommend synonyms or alternative phrases, augmenting the person writing. There are even dedicated AI tools that take writing advice one step further, such as Grammarly (#35).
- Augment video editors. A person in the midst of editing a video could receive a recommendation from an AI, suggesting appropriate music or cutting for a scene.
- Augment designers. A creative AI assistant can amplify innovative design processes by suggesting improvements in real-time. For instance, a car designer may receive suggestions for creating an appealing vehicle design or perhaps a more functional vehicle altogether.
- Augment corporate executives. Management can be empowered to run their business better, with an AI that suggests innovative ways to increase revenue. This is perhaps the ultimate form of augmentative AI: allowing the leadership of an organization to realize unknown improvements to the value chain.

As the most complicated and least adopted of the four strategies, companies investing in the innovation strategy can experience an enormous early advantage and a massive return on investment. Realizing the unexpected solutions that this strategy can provide is not easy, but worth attempting.

How To Use The Framework

The framework is an excellent approach to identifying, explaining, or understanding how AI could be used to empower any activity. Through one of the four strategies (efficiency, effectiveness, expert, and innovation) and ultimately one of the two actions (automation and augmentation), AI can empower just about anything.

Though I have focused on explaining the framework in the context of processes, one can naturally adopt any tool or solution onto the framework. However, doing so does create a context-sensitive result. For instance, if you were to map a solution such as Google Translate into the framework, you'd find that it can fit into different strategies depending on the process in which you seek to apply it. If you only need rough translations for personal use, you might use Google Translate as an autonomous tool – the efficiency strategy. If you need to communicate with someone who speaks a different language than you do, you may instead be using the AI to streamline communication, which would be the effectiveness strategy. And so on.

Now, let's get back to processes and discuss how to break processes down into subprocesses. If you are ever uncertain as to where a process fits into the framework, you are probably looking at a process too broadly, and need to split it up. I like to use the example of recruiting, as I think most of us have been on the receiving end of a recruitment process at least once.

Recruiting is a process that can be split into a large number of subprocesses. From the employers' point of view, examples of subprocesses could be creating a job ad, standing at a job fair, collecting resumes, analyzing said resumes, asking participants to perform online tests, inviting them to interviews, performing interviews, analyzing interviews, and so on and so forth. These steps may be improved with different AI strategies, and some may be cheaper or easier to improve than others. Rather than adopting AI to every subprocess along the way, one could (and should) opt to take it one subprocess at a time, using the framework.

Also, note that a process that fits into a specific strategy for a competitor may not fit into the same strategy for one's own business. For instance, the process of purchasing could fit into any of the four alternatives, all depending on the business model. For a grocery store, the purchasing of many goods can be automated, as many items in grocery stores are constant. As an example, AI can learn how much butter is being sold in each respective month and holiday, and autonomously order butter from wholesalers accordingly. A niche store, however, with shifting wares, may opt to adopt the expert strategy for the same process. They may use AI to find new products and negotiate deals for them. Thus, the process of purchasing items from a wholesaler could vary depending on the business model.

Next, it is crucial to note that the definition of "complex" is ever-changing. A process that was once considered to have a high data or work complexity may now be deemed to be simple. Over time, processes will move from top to bottom and from right to left, but likely not vice versa. For example, financial investment jobs are typically placed in the expert strategy, but there are already investment funds that are fully managed by an AI (#59), which would belong in the efficiency strategy. Similarly, educational and medical processes may move into the efficiency strategy, as AI could potentially come to execute such processes autonomously. There will likely be two separate forms of healthcare in the near future, one that is simple enough to be run entirely by AI, and one that requires human expertise.

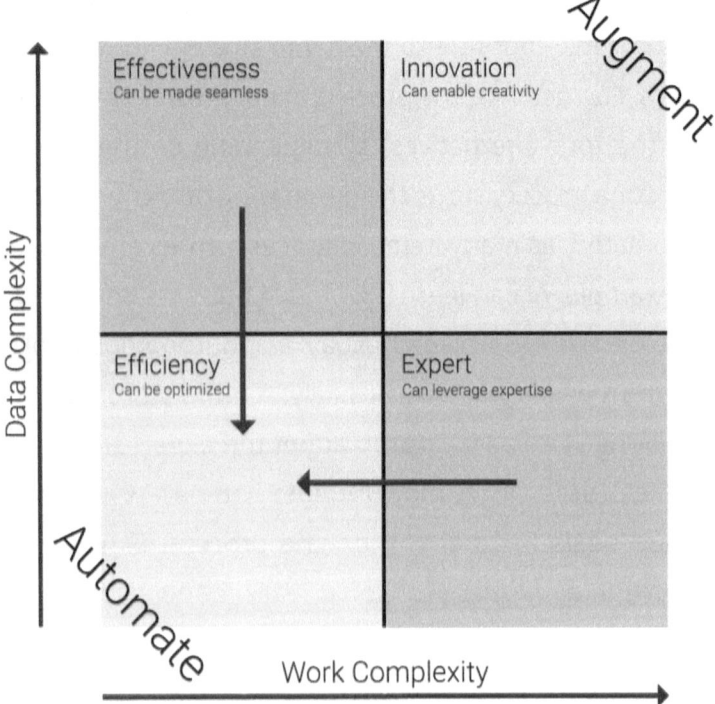

Companies can also reduce the work complexity of a process themselves. If a process lacks routines, they could create routines (or buy them), if they believe the process is better fit to be automated than augmented. Establishing routines is a lot more complicated than it sounds, however. Ask a friend of yours to explain their work in excruciating detail, and they might just find themselves unable to do so. A vast majority of jobs do *not* have *clearly* defined routines. They might *have* routines; they just can't *explain* them.

Regardless, companies who wish to explore AI will likely want to begin their AI journey on the left side of the framework. Strategies that are close to automation are often an easy first step to start an AI journey, though all organizations will be forced to invest in expert AI in the future.

As a final note on the placing of processes, I would like to bring up the need for human intimacy. Sometimes, a process has both low data and low work complexity, yet may still not be fit to be automated. An example of such a process could be to inform a person that they have developed cancer. Telling someone that they have received an illness can easily be accomplished by an AI, but it is a mentally daunting message to receive. It is best delivered by a human, don't you agree? I know – this is a bit of a blunt example, but there are many such processes where the need for a human connection cannot be understated. To name a case that's more down to earth, there are no doubt many high-end retail stores that would prefer not to replace human employees, as humans are perhaps able to provide better service to customers than machines are. In the future, it could even become prestigious to shop at stores with human clerks.

Transitioning the workforce

A massive shift is about to occur in the workforce. Today, many jobs are predictive and rules-based. Yet, as we've established, AI is better at performing such tasks than humans. Humans will, therefore, be replaced in many of these jobs. As AI can accomplish these tasks more efficiently, there will be more work that, in turn, requires human decision-making. This will result in more opportunities for complicated tasks for humans to process. As a consequence, the workforce will have to move from performing simple jobs to performing more complex jobs. Re-educating the workforce to handle these complicated tasks is very time-consuming. Therefore, the conclusion: companies will be forced to invest in expert AI that simplifies complicated work so that the workforce can more easily transition to performing them.

That last paragraph contained a lot of information to take in, didn't it? Let me give you a concrete example of this change. AI is both faster and more accurate at diagnosing patients than human doctors are, but humans are better at treating patients. Diagnosing is a predictive job while treating is a judging job. As AI can treat patients faster, doctors will have fewer patients in need of diagnosing but more patients in need of treatment. Thus: fewer jobs that require predicting, and more jobs that require judging. Nurses whose primary role previously may have been to diagnose patients could have their role changed into performing diagnoses instead, as AI might allow humans to self-diagnose. Expert systems help nurses and doctors perform the treatment of the illness that was discovered by the AI, augmenting their position.

Financial jobs are another example. Investors and accountants often have three to five years of University education, but what if expert AI could simplify these jobs to such a level that anybody could perform the same job after just a few months of training?

Augmentative AI is a necessary investment for organizations that have previously invested in automation. Ever since the first industrial revolution, humans have always been required to move to increasingly cognitively challenging jobs. This fourth industrial revolution is no exception. Augmentative AI can make those challenging jobs less challenging.

Understanding the stories

So, what do you think? I hope this chapter has made it easier to understand why and how people implement AI. Now, although the framework is best applied when examining individual processes, I would like to use this framework for a few trend analyses of the AI solutions that you read about in part two.

For the rest of part three, I'll cover some interesting trends that you may or may not have also noticed. These analyses are a simplification of reality.

AI Across The Organization

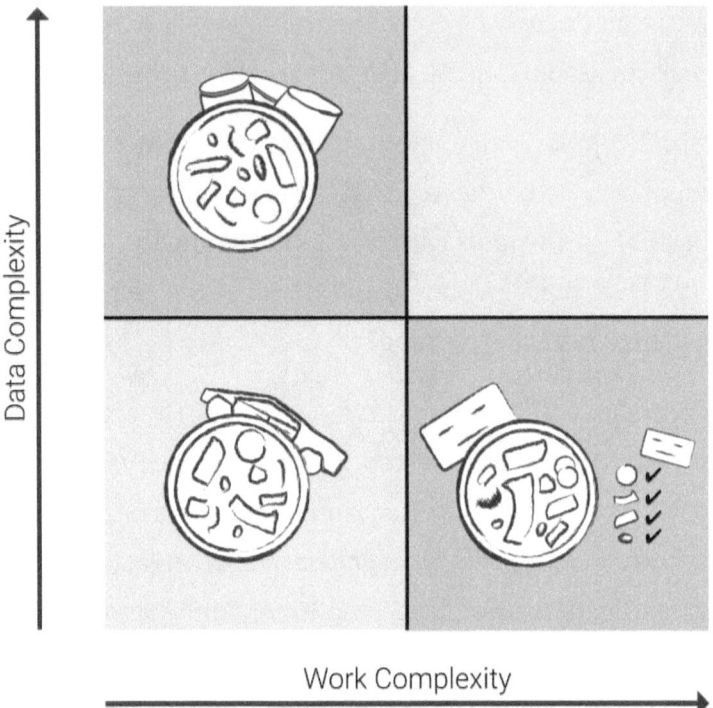

The international pizza chain Domino's Pizza might not be the first thing that comes to mind when one thinks of AI, but perhaps it should be. As you may have noticed while reading through part two, Domino's have been working towards utilizing AI across many layers of their business.

Indeed, Domino's have implemented at least three of the four AI strategies into their value chain.

Efficiency strategy: Domino's have been rolling out a pilot program for delivering pizzas with self-driving cars, with hopes of a full roll-out in the coming years (#44).

Effectiveness strategy: Customers can order a pizza from Domino's pizza chain through a natural language processing assistant, which Domino's has made available on platforms such as Google Home, Amazon Alexa, and Facebook Messenger (#45).

Expert strategy: Domino's takes a photo of every pizza the company's chefs cook. The picture is analyzed through image recognition AI to ensure that the quality is up to par and that the toppings are correct to order (#46).

Domino's adoption of AI is quite remarkable, mostly due to the fact that the company is in the category of companies *least* likely to adopt AI. It's a multinational non-tech company founded in the 1960s with a streamlined value chain and a dominating position on the market. That's precisely the type of company that is least likely to realize AI's potential. For large companies with optimized value chains, adopting AI can be a colossal challenge, as processes need to be fundamentally reworked.

Yet, Domino's have carefully rolled out AI initiatives and trials without replacing old processes entirely. They have adopted AI onto various activities in different regions, worked across their entire business, and realized the value of selecting the right AI strategy for the right business process.

This approach has allowed the company to stay ahead of the curve and to remain innovative without having to disrupt its entire value chain.

Similar Products, Different Strategies

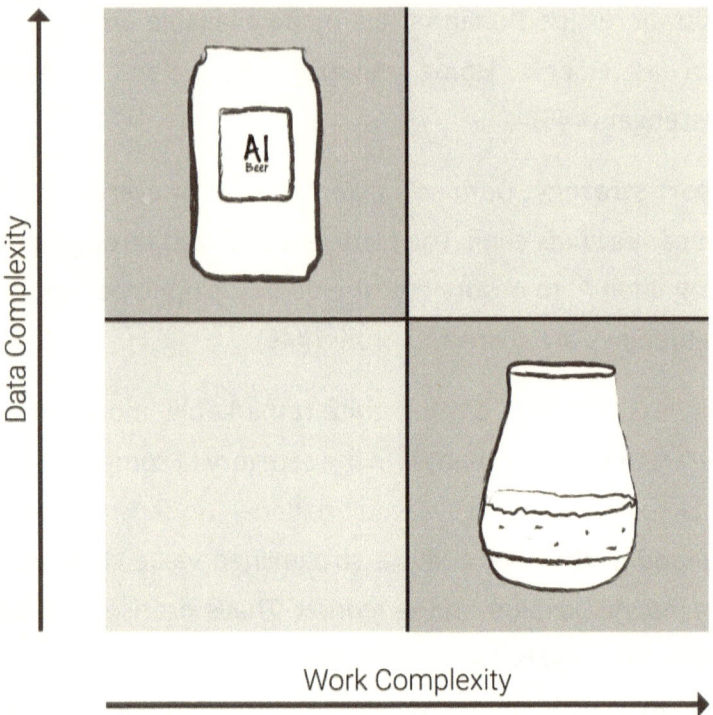

I mentioned that different companies might adopt different AI strategies for the same processes (such as purchasing from wholesalers) depending on their business model. Well, do you remember IntelligentX (#5) and Mackmyra (#6)? The former creates beer using AI, while the latter uses AI to craft whiskey. Both companies are creating alcoholic beverages using AI, yet you may have noticed that their business models are entirely different.

IntelligentX adopted the effectiveness strategy. Their AI is built to interact with their customers – asking them questions to discern their palette. Using the results provided by the AI, the company creates customized beer for each of their customers. Furthermore, IntelligentX appears to have built its entire business around AI, as they don't offer any products that aren't AI-powered. Their sole competitive advantage seems to lay in their AI.

Mackmyra, meanwhile, adopted the expert strategy. Their AI is used in collaboration with human experts in order to create high-quality whiskey. Mackmyra does not create customized whiskeys for each of its customers. They created one single blend, through human-machine collaboration. Mackmyra also offers plenty of non-AI-powered products, displaying a more conventional business model with a large selection of traditionally developed products.

Obviously, comparing beer with whiskey is like comparing apples with oranges, but I wanted to exemplify how AI can be adopted in fundamentally different ways, even within product development. IntelligentX could have chosen the same approach as Mackmyra but opted differently.

Again: AI can be used to either automate or augment processes using one of four strategies. The first strategy that comes to mind for a particular business might not necessarily be the best strategy to adopt. Just because a competitor is utilizing AI in a certain way, doesn't mean that your business should as well.

Understanding the strengths of each AI strategy is vital.

AI Masters Games

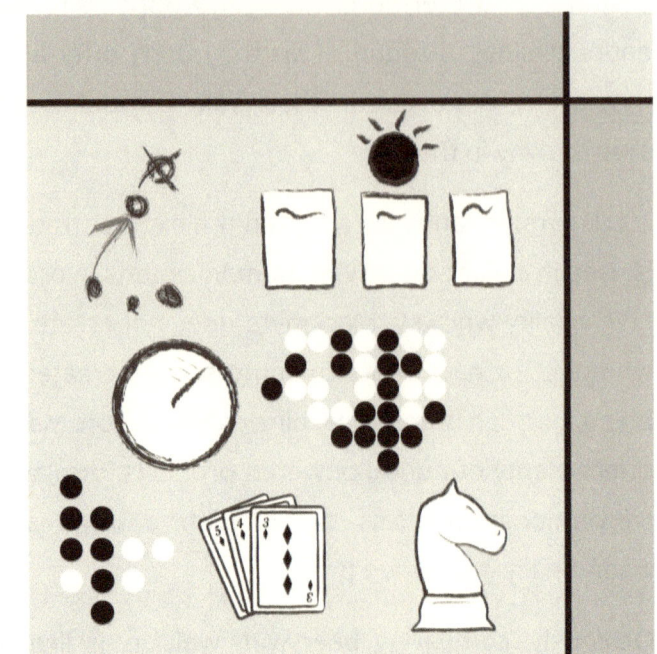

Data Comple

Work Comple

Be it board games or video games – AI has proven to consistently outperform humans at playing games. Starting with AI's victory versus the world champions in chess (#13) and Othello (#14) in 1997, AI has gone on to defeat the best of the best in Jeopardy (#16) and Go (#15). AI will soon beat the best players of StarCraft 2 (#18) and Poker (#3) as well. In single-player games such as Super Mario World (#17), AI has also proven to be able to learn to play the game to perfection.

Mastering games might not sound like a big deal. After all, no one is going to lose their job because an AI learned to play chess incredibly well, are they? And humans certainly haven't stopped competing in Jeopardy just because an AI defeated the champions. No, you should think of these events as technological showcases. These are astounding displays of automation and self-learning AI. Many of these games are incredibly complicated. The powerful technology displayed here could, will, and already has been applied to more complex processes that do affect day-to-day roles.

Go is likely over 3,000 years old. Can you believe that? This game has existed for many millennia. After so many years, most people were certain that the game had been mastered, that all possible ways to play the game had been figured out. Yet, DeepMind's AI was able to discover entirely new strategies to the game. When DeepMind's AlphaGo played a five-game Go match against the world champion, Lee Sedol, it stunned both the champion and the audience by playing a move never seen before. It was a move that would be analyzed for years.

This is a quality of AI that most people may not realize. AI is capable of discovering methods that humans have never even thought of before. Not just in games, but in other activities as well.

Jeopardy, then, is a complicated trivia game in which answers are the questions, and the questions are the answers. In this game, the challenge for AI was to comprehend the question. What is the question (answer) to "a thief, or a bent part of an arm." It's "what is a crook?" but IBM's Watson didn't know that. Despite getting some answers wrong, Watson ended up defeating two of the all-time best Jeopardy contestants, with a significant margin. This was very much a technological showcase for IBM – in fact, if we are going to be frank and call it for what it is, it's an advertisement – but it's a legitimate one, for the technology truly is remarkable. The AI displayed an exceptional level of understanding of the human language.

Next, let's briefly discuss Poker and StarCraft 2. The key challenge for AI in mastering both of these games is that they contain elements hidden to the player. See, in games like chess, Othello, and Go, all game pieces are laid out on the board. In Poker, however, the players cannot see their opponent's cards. In StarCraft 2, the players cannot see their opponent's units. And yet AI was able to overcome these challenges through powerful algorithms that can make calculations even on hidden information.

For all of the examples named above, a solitary AI was taught to play a single specific game as well as possible. This means that the AI that mastered chess doesn't know how to play Othello, and vice versa, which perhaps significantly bleaks the prowess of AI. Fine, Logistello defeated the world champion in Othello, but ask the AI to play chess against an eight-year-old and Logistello wouldn't even know how to place a single piece. Clearly, these AI are modular pieces of technology that are no match to the more capable brains of humans.

Now is the time we discuss DeepMind's MuZero. This quite recent AI has, at the time of writing, mastered over 50 games. This list of games includes chess, Go, Shogi, various Atari games, and a lot more. It learned to play these games without having any knowledge of the games' rules beforehand. [178]

Yes, every expectation for AI has been shattered. Before AI mastered Jeopardy, most people confidently said that it couldn't be done. Before AI mastered Go, most people confidently said that it couldn't be done. And until very recently, most people confidently said that AI would always remain modular; case-specific.

On a final note, it must also be said that these AI are all working in very restricted environments. All of these games have clearly defined rules and boundaries – which real-life perhaps does not. This makes them ideal settings for AI to flourish. There's a list of every possible move in chess, but there's no list of every possible move in business, in art, in life.

Not yet.

AI Understands Human Language

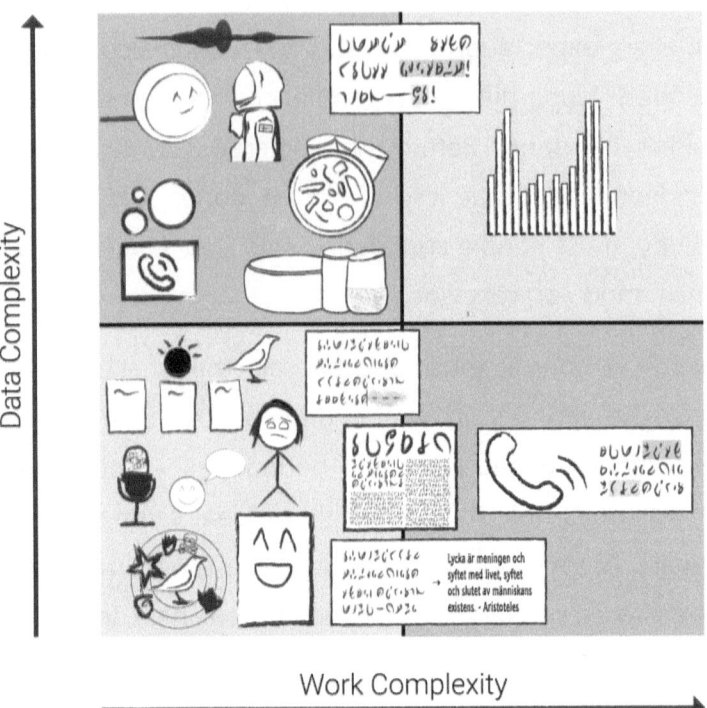

Data Complexity

Work Complexity

The field of natural language processing is exploding! Did you notice just how many of the stories in this book involve natural language processing to some extent? For every day that passes, AI is getting better at listening, comprehending, and speaking. People are using their voice or human-language text to perform all kinds of actions on the devices. The way we think about user interfaces is changing as companies move towards implementing digital assistants that users can talk or chat with.

Most of the solutions involving natural language processing are located in the leftmost part of the framework. This is because NLP is typically used to either optimize workflows or to streamline communication. However, as NLP is becoming increasingly sophisticated, it is transitioning into empowering expert roles. For example, writing a book, such as this one, would be an example of a process often placed in the rightmost part of the framework, given its high work complexity. And the very book you are reading right now has indeed been written together with several AI's.

Anyway, let's look at the efficiency strategy first, which is in the bottom-left corner of the framework. These are autonomous AI solutions that have optimized processes. There's Jeopardy-playing Watson (#16), the rhetoric-loving Project Debater (#30), the article-transforming-podcast AI Playpost (#28), the heatmap-making police tool that reads social media (#69), the bully-detecting Instagram AI (#52), the description-writing AI (#41), the therapy-providing AI (#56) and Microsoft's short-lived racist chatbot Tay (#26). One way or another, all of these AI autonomously execute tasks of various importance.

In the effectiveness strategy, which is in the upper-left corner of the framework, AI is used to make people more effective, primarily by simplifying communication between humans. There's CIMON, the astronaut-bot that lived on the ISS (#27), Domino's pizza-order-taking chatbot (#45), and Google's impressive appointment-booking AI Duplex (#29), along with the more general digital assistants Siri (#24) and home assistants (#25).

In the expert strategy, the real-time conversation-analyzing AI Cogito is improving the performance of experts (#76), while the song-sampling AI that gave us Blue Jeans and Bloody Tears (#34) sits in the innovation strategy, having empowered the creativity of music composers.

Some solutions overlap numerous strategies. This is due to the fact that these AI can be deployed for a wide range of processes, meaning the context that the tool is applied to alters its value-creation strategy. This is probably true for several of the other solutions that I arrogantly and confidently placed within specific frameworks as well.

Crossing between the efficiency and expert strategies is both the news-writing AI from RADAR (#75) and the well-known translation AI Google Translate (#63). The former is capable of fully autonomously writing news stories, though it is typically used as an expert tool, with human experts applying their writing skills on top of the writing produced by the AI. In this latter scenario, the product of the AI is seen as a draft rather than a finished product. As for Google Translate, the context in which the solution is used changes the strategy applied for it. An autonomous tool for translating entire websites or an expert tool for translating individual sentences in a broader context? It depends on the use case.

Crossing over the effectiveness and innovation strategies is Grammarly (#35), the AI tool that analyzes text and suggests improvements. Depending on the process to which the tool is applied, it could be seen as a way to make communication more effective or as a way of empowering the creative art of writing. An example process of the former scenario could be a writer who explains concepts through a vocabulary that is too complex for its target audience. This writer may request the AI to correct them if any word they use is too complicated, thus ensuring that the text is appropriately communicated. An example process of the latter scenario could be a creative writer in the midst of writing a fictitious novel, who uses the AI to discover romantic synonyms that elevate the text.

There are even more stories from part two that could arguably be added to this trend, such as Instagram's emoji-learning AI (#82), though it's debatable.

As NLP has become rather competent in recent years, so too has its popularity skyrocketed. Go to virtually any website, and a little window will pop-up in a corner saying something along the lines of "let me know if there's anything I can do for you."

I will end this trend with a prediction. I predict that digital assistants will become as integrated into your daily life as smartphones are, if not even more. AI will learn your political orientation, your romantic preferences, and your career desires. It will become the norm to ask your AI whether to vote for or against the proposition, whether to date Robert or Brad, and whether or not to accept the job offer. The notion that people would ask a machine if a person is the love of their life might sound like nonsensical sci-fi-shenanigans, but given the speed of AI advancement, I believe that this will happen within our lifetime.

AI Enables Product Development

Another emerging trend in AI that really entices me personally is the usage of AI for product development. It is a lot more common than one might be led to believe. Besides being incredible at playing games, understanding speech, and automating value chains, AI is also a great companion for building actual products. This is sometimes done autonomously, but often through human-machine collaboration, with AI acting either as an assistant for experts or as a tool for creatives.

Whenever I argue for AI's ability to assist with product development to clients, I am often initially met with a moment of skepticism. But it has been proven that AI is already very capable in this regard. The world's most effective flu vaccine was created by an AI (#2), and AI has assisted, or is continuing to assist, in the creation of beer (#5), whiskey (#6), food recipes (#8), food products (#9), perfumes (#10), and software (#100). Note that I placed the last story in-between the expert and innovation strategies simply because I made that story quite broad, with a rather wide variety of solutions. Most of the examples of AI in story #100 are expert-oriented, but some would perhaps belong in the innovation strategy, such as the design-empowering AI Sketch2Code.

I predict that this trend will increase considerably. There are two paths for companies to take. The first option is to use AI to create products "on the fly" that are customized for each individual consumer – without the need for much human input from the company's side. Crafting unique products that are made specifically for each customer profile is a wonderful approach for bringing digital expectations to the physical world. The second option is to use AI to empower human product developers. AI can learn the components that make up their products and elevate them further.

Take the battle between Walmart and Amazon. The offline commerce giant versus the online commerce giant. These companies are competing for the same customers. As prices are starting to become close to as low as they can be, Walmart is inventing brand new food products in order to claim (or perhaps *reclaim*) customers from Amazon. [179]

AI Enables Creativity

While AI is certainly capable of producing creative solutions all by itself, it is often used as a creative assistant for humans. Though sometimes hard to monetize, this is nonetheless a rising trend in AI usage.

Personally, I am very passionate about the usage of AI to revolutionize the school system. AI can create personalized course plans and innovative ways of teaching that caters to the needs of each individual pupil. Though this form of AI is difficult to develop and still in early stages, millions of Chinese pupils are already being taught by an AI, for instance (#1).

Music and art are two obvious areas for AI use as well. Though I chose to only include one example from each of these categories, there are surely hundreds of examples of AI being used to either autonomously create art and music by itself or for augmenting human creatives. So far, AI in the latter category has been more successful than the former, though AI is becoming better and better at creating creative content autonomously. The two examples I included in the book both belong to the latter category.

The song I included in the book, Blue Jeans and Bloody Tears (#34), was made with human-machine collaboration. The AI provided samples for human creatives to mix. The painting (#97), which, as you might recall, was sold for 432,000 USD (I still find this to be crazy), was also made in large parts with human-machine collaboration. It is unknown just how much of the painting was made autonomously.

In previous trends, I covered the two stories of the writing-enhancing AI Grammarly (#35) and the group of developer-empower AI's (#100), which I also included here.

Though this book focuses on the present and near-future, please allow me to delve into a sci-fi vision for a moment. Creativity-enabling AI is the final AI. It will aid humanity in finding meaning. Today, many humans find meaning in their work. In a possible future, wherein all boring jobs have been automated and the state guarantees everyone a basic income thanks to corporate costs being minimal, creativity-enabling AI may be what allows humans to find purpose in life.

AFTERWORD

Stay In The Race

Thank you for reading This Is Real AI! This first book of mine was supposed to be a small little project, in which I basically converted one of my websites[4] into a book. I naively thought it wouldn't be much work. Of course, it ended up being a huge undertaking, but it has been a whole lot of fun!

Whether a newcomer to artificial intelligence or a seasoned veteran, I hope you have enjoyed reading this book. I hope that it has provided you with inspiration and a deeper understanding and curiosity for AI. If you have any questions or any feedback, or if you simply wish to continue the discussion, please feel free to reach out to me via e-mail[5] or through my LinkedIn[6]. I'd be delighted to hear your thoughts on the book, on AI, or just the future in general. If you enjoyed the book, please do consider leaving a book review wherever you purchased it. Book reviews help a lot!

Now, before I leave you, I would like to suggest some alternatives for continued learning.

For starters, if you'd like to stay up to date with the latest news in AI, then I do have a couple of tips. Say what you will about Reddit, but for the dedicated, I do consider the Reddit subcommunity r/artificial[7] to be a great place to keep track of everything that's happening in AI. For the person who wants

---

4 https://thisisrealai.com
5 thisisrealai@gmail.com
6 https://www.linkedin.com/in/jacobbergdahl/
7 https://www.reddit.com/r/artificial/

the gems rather than every nitty little piece of news, AI News[8] is an excellent resource for following the latest in AI. If you'd rather just get the big news, bundled with other general tech news, platforms such as The Verge[9], Tech Crunch[10], and Medium[11] are fantastic places to keep track of the biggest news in AI, as well as expert opinions on the subject. These are all websites that I visit daily.

Speaking of me, I write quite a bit about AI and tech. You can follow me on LinkedIn[12] or Medium[13] if you're interested in more of my thoughts on artificial intelligence and other topics.

There are also plenty of fantastic books on artificial intelligence. My personal favorite is Max Tegmark's *Life 3.0*. In it, Tegmark discusses the path to and the consequences of artificial general intelligence. The book has a great blend of storytelling and facts. It's a really good read!

Pedro Domingo's *The Master Algorithm* is another outstanding book. Domingo discusses the journey towards discovering the final algorithm – the one AI that can solve virtually any issue. Domingo walks the reader through the five main schools of machine learning, while also including some interesting business scenarios, should you be interested in finding more business cases. Domingo's book is in parts slightly more

---

[8] https://artificialintelligence-news.com/
[9] https://www.theverge.com/
[10] https://techcrunch.com/
[11] https://medium.com/
[12] https://www.linkedin.com/in/jacobbergdahl/
[13] https://medium.com/@jacobbergdahl_47336

complicated than Tegmark's piece, though you can simply skip the complicated sections if you prefer, and you'd still be able to follow along.

Finally, Yuval Noah Harari's book *Homo Deus* is an excellent choice for those who seek a much broader, philosophical overview of the future. Harari's book is not specific to AI, instead covering the history and future of humanity in general. Every single page in *Homo Deus* is thrilling. Reading the book is like going on a journey of discovery about the very foundation of what makes us human. I love this book.

If you do want to get more technical, then check out Andrew Ng's online course on machine learning[14]. It's one of the most renowned machine learning courses in the world. Enrolling is free of charge.

Try also to use the framework that I presented in part three. Again, it's an incredibly useful and robust framework that's simple to use. If you're interested in getting started with implementing AI solutions yourself, take the inspirations you've gained from part two with you, evaluate your own business, and reach out to consultants who can help you.

Once more, from the bottom of my heart, thank you for reading through this book.

Welcome to the world of AI.

---

[14] https://www.coursera.org/learn/machine-learning

References

[1]   T. Dutton, "An Overview of National AI Strategies,"
        Politics + AI, 28 06 2018. [Online]. Available:
        https://medium.com/politics-ai/an-overview-of-national-
        ai-strategies-2a70ec6edfd. [Accessed 08 12 2019].

[2]   J. Vincent, "Putin says the nation that leads in AI 'will be
        the ruler of the world'," The Verge, 04 09 2017. [Online].
        Available:
        https://www.theverge.com/2017/9/4/16251226/russia-
        ai-putin-rule-the-world. [Accessed 08 12 2019].

[3]   M. Hutson, "Trump to launch artificial intelligence
        initiative, but many details lacking," Science, 11 02 2019.
        [Online]. Available:
        https://www.sciencemag.org/news/2019/02/trump-
        launch-artificial-intelligence-initiative-many-details-
        lacking. [Accessed 08 12 2019].

[4]   Z. Xin and C. Chi-yuk, "Develop and control: Xi Jinping
        urges China to use artificial intelligence in race for tech
        future," South China Morning Post, 31 10 2018. [Online].
        Available: https://www.scmp.com/economy/china-
        economy/article/2171102/develop-and-control-xi-
        jinping-urges-china-use-artificial. [Accessed 08 12 2019].

[5]   K. Costello, "Gartner Survey Shows 37 Percent of
        Organizations Have Implemented AI in Some Form,"
        Gartner, 21 01 2019. [Online]. Available:

https://www.gartner.com/en/newsroom/press-releases/2019-01-21-gartner-survey-shows-37-percent-of-organizations-have. [Accessed 15 12 2019].

[6]   M. Purdy and P. Daugherty, "How AI Boosts Industry Profits And Innovation," Accenture, 2017. [Online]. Available: https://www.accenture.com/fr-fr/_acnmedia/36dc7f76eab444cab6a7f44017cc3997.pdf. [Accessed 15 12 2019].

[7]   Accenture, "Future Workforce Survey - Banking - Realizing the Full Value of AI," Accenture, [Online]. Available: https://www.accenture.com/_acnmedia/PDF-77/Accenture-Workforce-Banking-Survey-Report#zoom=50. [Accessed 15 12 2019].

[8]   PWC, "Consumer Intelligence Series - Bot.Me: A revolutionary partnership," PWC, [Online]. Available: http://pwcartificialintelligence.com/. [Accessed 15 12 2019].

[9]   S. Ransbotham, D. Kiron, P. Gerbert and M. Reeves, "Reshaping Business With Artificial Intelligence," MIT Sloan Management Review, 06 09 2017. [Online]. Available: http://sloanreview.mit.edu/projects/reshaping-business-with-artificial-intelligence/. [Accessed 15 12 2019].

[10]   Sage, "Survey Reveals Nearly 50% of Consumers State They 'Have No Idea What Artificial Intelligence is About'," Sage, 07 11 2017. [Online]. Available:

https://www.sage.com/en-us/news/press-releases/2017/11/survey-reveals-nearly-50-percent-of-consumers-have-no-idea-what-artificial-intelligence-is-about/. [Accessed 15 12 2019].

[11]   S. K. Robinson, "Do schools kill creativity?," TED, 07 01 2007. [Online]. Available: https://www.youtube.com/watch?v=iG9CE55wbtY. [Accessed 13 10 2019].

[12]   Squirrel AI, "Squirrel AI," [Online]. Available: http://squirrelai.com/. [Accessed 13 10 2019].

[13]   C. K. Ho, "AI education unicorn Squirrel targets foreign markets with plans for mathematics, Mandarin lessons," South China Morning Post, 15 07 2019. [Online]. Available: https://www.scmp.com/tech/start-ups/article/3018297/chinese-education-unicorn-squirrel-ai-targets-foreign-markets-plans. [Accessed 13 10 2019].

[14]   H. Masige, "Australian Researchers Have Just Released The World's First AI-Developed Vaccine," Science Alert, 13 07 2019. [Online]. Available: https://www.sciencealert.com/the-world-s-first-ai-developed-vaccine-could-prevent-another-horror-flu-season. [Accessed 13 10 2019].

[15]   J. Vincent, "Facebook and CMU's 'superhuman' poker AI beats human pros," The Verge, 11 07 2019. [Online]. Available: https://www.theverge.com/2019/7/11/20690078/ai-

poker-pluribus-facebook-cmu-texas-hold-em-six-player-no-limit. [Accessed 13 10 2019].

[16]   N. Garun, "A Chinese AI startup is tracking lost dogs using their nose prints," 13 07 2019. [Online]. Available: https://www.theverge.com/2019/7/13/20693064/megvii-chinese-ai-facial-recognition-lost-pets-dogs-cats-surveillance. [Accessed 13 10 2019].

[17]   IntelligentX, "How it works," 2019. [Online]. Available: https://join.intelligentx.ai/pages/how-it-works. [Accessed 13 10 2019].

[18]   E. Dedezade, "Meet the world's first AI-created whisky," Microsoft, 13 05 2019. [Online]. Available: https://news.microsoft.com/europe/features/meet-the-worlds-first-ai-created-whisky/. [Accessed 13 10 2019].

[19]   Mackmyra, "Intelligens," Mackmyra, [Online]. Available: https://mackmyra.com/product/intelligens/. [Accessed 13 10 2019].

[20]   P. Domingos, The Master Algorithm: How the Quest for the Ultimate Learning Machine Will Remake Our World, Basic Books, 2017.

[21]   B. Anderson, "The Rise of the Weaponized AI Propaganda Machine," Scout, 13 02 2017. [Online]. Available: https://medium.com/join-scout/the-rise-of-the-weaponized-ai-propaganda-machine-86dac61668b. [Accessed 13 10 2019].

[22]   Tovo Labs, "Tovo Labs," 2019. [Online]. Available:
       https://tovolabs.com/. [Accessed 13 10 2019].

[23]   R. Brandt, "Chef Watson has arrived and is ready to help
       you cook," IBM, 01 01 2016. [Online]. Available:
       https://www.ibm.com/blogs/watson/2016/01/chef-
       watson-has-arrived-and-is-ready-to-help-you-cook/.
       [Accessed 13 10 2019].

[24]   R. Lougee, "Using AI to Develop New Flavor
       Experiences," IBM, 05 02 2019. [Online]. Available:
       https://www.ibm.com/blogs/research/2019/02/ai-new-
       flavor-experiences/. [Accessed 13 10 2019].

[25]   Symrise, "Breaking new fragrance ground with Artificial
       Intelligence (AI): IBM Research and Symrise are working
       together," Symrise, 24 10 2018. [Online]. Available:
       https://www.symrise.com/newsroom/article/breaking-
       new-fragrance-ground-with-artificial-intelligence-ai-ibm-
       research-and-symrise-are-workin/. [Accessed 15 10
       2019].

[26]   B. Osterath, "Artificial intelligence creates perfumes
       without being able to smell them," DW, 31 05 2019.
       [Online]. Available: https://www.dw.com/en/artificial-
       intelligence-creates-perfumes-without-being-able-to-
       smell-them/a-48989202. [Accessed 15 10 2019].

[27]   B. Strope, "Efficient Smart Reply, now for Gmail," Google
       AI Blog, 17 05 2017. [Online]. Available:

https://ai.googleblog.com/2017/05/efficient-smart-reply-now-for-gmail.html. [Accessed 20 10 2019].

[28]   J. Jersin, "How LinkedIn Uses Automation and AI to Power Recruiting Tools," LinkedIn Talent Blog, 10 10 2017. [Online]. Available: https://business.linkedin.com/talent-solutions/blog/product-updates/2017/how-linkedin-uses-automation-and-ai-to-power-recruiting-tools. [Accessed 20 10 2019].

[29]   M. McPhee, K. Baker and C. Siemaszko, "Deep Blue, IBM's supercomputer, defeats chess champion Garry Kasparov in 1997," Daily News, 10 05 2015. [Online]. Available: https://www.nydailynews.com/news/world/kasparov-deep-blues-losingchess-champ-rooke-article-1.762264. [Accessed 20 10 2019].

[30]   The Associated Press, "Computer Beats Champ Again - This Time in Othello," New York Times, 09 08 1997. [Online]. Available: https://archive.nytimes.com/www.nytimes.com/library/cyber/week/080997othello.html. [Accessed 20 10 2019].

[31]   DeepMind, "AlphaGo," DeepMind, [Online]. Available: https://deepmind.com/research/case-studies/alphago-the-story-so-far. [Accessed 20 10 2019].

[32]   J. Best, "IBM Watson: The inside story of how the Jeopardy-winning supercomputer was born, and what it

wants to do next," Tech Republic, 09 09 2013. [Online]. Available: https://www.techrepublic.com/article/ibm-watson-the-inside-story-of-how-the-jeopardy-winning-supercomputer-was-born-and-what-it-wants-to-do-next/. [Accessed 20 10 2019].

[33] SethBling, "MarI/O - Machine Learning for Video Games," YouTube, 13 06 2015. [Online]. Available: https://www.youtube.com/watch?v=qv6UVOQ0F44. [Accessed 20 10 2019].

[34] P. Sun, X. Suna and L. Hana, "TStarBots: Defeating the Cheating Level Builtin AI inStarCraft II in the Full Game," Tencent, 27 12 2018. [Online]. Available: https://arxiv.org/pdf/1809.07193.pdf. [Accessed 20 10 2019].

[35] N. Statt, "DeepMind's StarCraft 2 AI is now better than 99.8 percent of all human players," The Verge, 30 10 2019. [Online]. Available: https://www.theverge.com/2019/10/30/20939147/deepmind-google-alphastar-starcraft-2-research-grandmaster-level. [Accessed 14 11 2019].

[36] Lumen5, "Lumen5," Lumen5, [Online]. Available: https://lumen5.com/. [Accessed 25 10 2019].

[37] R. Gurion, "Making Uber for Business Better for Every Business, Everywhere," Uber Newsroom, 13 08 2018. [Online]. Available: https://www.uber.com/newsroom/making-uber-

business-better-every-business-everywhere/. [Accessed 25 10 2019].

[38]   R. Waliany, L. Kang, E. Murati, M. S. Amin and N. Volk, "How Trip Inferences and Machine Learning Optimize Delivery Times on Uber Eats," Uber Engineering, 15 06 2018. [Online]. Available: https://eng.uber.com/uber-eats-trip-optimization/. [Accessed 25 10 2019].

[39]   J. Bird, "Chilling Or Thrilling? JD.com Founder Envisions A '100%' Robot Workforce," Forbes, [Online]. Available: https://www.forbes.com/sites/jonbird1/2018/04/27/chilling-or-thrilling-jd-coms-robotic-retail-future/. [Accessed 25 10 2019].

[40]   JD.com, "JD's Unmanned Store Goes International," JD.com Corporate Blog, 02 08 2018. [Online]. Available: https://jdcorporateblog.com/jds-unmanned-store-goes-international/. [Accessed 25 10 2019].

[41]   H. Murayama, "China's unmanned store boom ends as quickly as it began," Nikkei, 17 06 2019. [Online]. Available: https://asia.nikkei.com/Business/Business-trends/China-s-unmanned-store-boom-ends-as-quickly-as-it-began. [Accessed 26 10 2019].

[42]   Alibaba, "ET City Brain," Alibaba Cloud, [Online]. Available: https://www.alibabacloud.com/et/city. [Accessed 26 10 2019].

[43]   the unofficial AppleKeynotes channel, "YouTube," the unofficial AppleKeynotes channel, 22 03 2013. [Online].

Available:
https://www.youtube.com/watch?v=agzltTz35QQ.
[Accessed 26 10 2019].

[44] Vox Creative, "A brief history of voice assistants," The
Verge, 13 09 2018. [Online]. Available:
https://www.theverge.com/ad/17855294/a-brief-history-
of-voice-assistants. [Accessed 11 11 2019].

[45] B. Kinsella, "Amazon Announces 80,000 Alexa Skills
Worldwide and Jeff Bezos Earnings Release Quote
Focuses Solely on Alexa Momentum," Voicebot, 31 01
2019. [Online]. Available:
https://voicebot.ai/2019/01/31/amazon-announces-
80000-alexa-skills-worldwide-and-jeff-bezos-earnings-
release-quote-focuses-solely-on-alexa-momentum/.
[Accessed 11 11 2019].

[46] E. Kim, "Amazon Echo owners spend more on Amazon
than Prime members, report says," CNBC, 03 01 2018.
[Online]. Available:
https://www.cnbc.com/2018/01/03/amazon-echo-
owners-spend-more-on-amazon-than-prime-
members.html. [Accessed 11 11 2019].

[47] G. Sterling, "Alexa devices maintain 70% market share in
U.S. according to survey," Marketing Land, 09 08 2019.
[Online]. Available: https://marketingland.com/alexa-
devices-maintain-70-market-share-in-u-s-according-to-
survey-265180. [Accessed 11 11 2019].

[48]   E. Hunt, " Tay, Microsoft's AI chatbot, gets a crash course in racism from Twitter," The Guardian, 24 03 2016. [Online]. Available: https://www.theguardian.com/technology/2016/mar/24/ tay-microsofts-ai-chatbot-gets-a-crash-course-in-racism- from-twitter?CMP=twt_a-technology_b-gdntech. [Accessed 14 11 2019].

[49]   C. Davenport, "SpaceX is flying an artificially intelligent robot named CIMON to the International Space Station," Washington Post, 29 06 2018. [Online]. Available: https://www.washingtonpost.com/news/the- switch/wp/2018/06/29/spacex-is-flying-an-artificially- intelligent-robot-named-cimon-to-the-international- space-station/. [Accessed 14 11 2019].

[50]   Airbus, "CIMON is back on Earth after 14 months on the ISS," Airbus, 28 08 2019. [Online]. Available: https://www.airbus.com/newsroom/press- releases/en/2019/08/cimon-is-back-on-earth-after-14- months-on-the-iss.html. [Accessed 14 11 2019].

[51]   Playpost, "Instant podcasts of every article," Playpost, 2019. [Online]. Available: https://playpost.app/. [Accessed 14 11 2019].

[52]   J. Callaham, "What is Google Duplex and how do you use it?," Android Authority, 18 10 2019. [Online]. Available: https://www.androidauthority.com/what-is-google- duplex-869476/. [Accessed 13 11 2019].

[53]   N. Garun, "One year later, restaurants are still confused by Google Duplex," The Verge, 09 05 2019. [Online]. Available: https://www.theverge.com/2019/5/9/18538194/google-duplex-ai-restaurants-experiences-review-robocalls. [Accessed 13 11 2019].

[54]   IBM, "7152: Pushing the Frontiers of AI: Project Debater," IBM, 2019. [Online]. Available: https://www.ibm.com/events/think/watch/replay/120118800/. [Accessed 14 11 2019].

[55]   IBM, "How does Project Debater work?," IBM, 2019. [Online]. Available: https://www.research.ibm.com/artificial-intelligence/project-debater/how-it-works/. [Accessed 14 11 2019].

[56]   Carnegie Mellon University, "Navlab: The Carnegie Mellon University Navigation Laboratory," Carnegie Mellon University, [Online]. Available: https://www.cs.cmu.edu/afs/cs/project/alv/www/index.html. [Accessed 16 11 2019].

[57]   R. Schmelzer, "What Happens When Self-Driving Cars Kill People?," Forbes, 26 09 2019. [Online]. Available: https://www.forbes.com/sites/cognitiveworld/2019/09/26/what-happens-with-self-driving-cars-kill-people/. [Accessed 16 11 2019].

[58]   SAE International, "AUTOMATED DRIVING: LEVELS OF
        DRIVING AUTOMATION ARE DEFINED IN NEW SAE
        INTERNATIONAL STANDARD J3016," 2014. [Online].
        Available:
        https://cdn.oemoffhighway.com/files/base/acbm/ooh/d
        ocument/2016/03/automated_driving.pdf. [Accessed 16
        11 2019].

[59]   A. J. Hawkins, "Tesla's 'Full Self-Driving' feature may get
        early-access release by the end of 2019," The Verge, 23
        10 2019. [Online]. Available:
        https://www.theverge.com/2019/10/23/20929529/tesla-
        full-self-driving-release-2019-beta. [Accessed 16 11
        2019].

[60]   M. Bergen, "Google's Secret 'Trashy Video' AI Cleans Up
        YouTube Homepage," Bloomberg, 23 05 2019. [Online].
        Available:
        https://www.bloomberg.com/news/articles/2019-05-
        23/google-s-secret-trashy-video-ai-cleans-up-youtube-
        homepage. [Accessed 16 11 2019].

[61]   The Drum, "Nutella: Nutella Unica by Ogilvy Italy," The
        Drum: Creative Works, 06 2017. [Online]. Available:
        https://www.thedrum.com/creative-
        works/project/ogilvy-italy-nutella-nutella-unica.
        [Accessed 16 11 2019].

[62]   Sweaty Machines, "a Eurovision song created by Artificial
        Intelligence: Blue Jeans and Bloody Tears," Sweaty
        Machines, 13 05 2019. [Online]. Available:

https://www.youtube.com/watch?v=4MKAf6YX_7M.
[Accessed 16 11 2019].

[63]   Grammarly, "How We Use AI to Enhance Your Writing |
         Grammarly Spotlight," Grammarly, 14 08 2018. [Online].
         Available: https://www.grammarly.com/blog/how-
         grammarly-uses-ai/. [Accessed 16 11 2019].

[64]   A. Mosseri, "Working to Stop Misinformation and False
         News," Facebook, 07 04 2017. [Online]. Available:
         https://www.facebook.com/facebookmedia/blog/workin
         g-to-stop-misinformation-and-false-news. [Accessed 16
         11 2019].

[65]   BuzzFeedVideo, "You Won't Believe What Obama Says In
         This Video! 🙂," BuzzFeed, 17 04 2018. [Online].
         Available:
         https://www.youtube.com/watch?v=cQ54GDm1eL0.
         [Accessed 16 11 2019].

[66]   Future Advocacy, "We're researching the potential
         impacts of deepfakes on society," Future Advocacy,
         [Online]. Available:
         http://futureadvocacy.com/deepfakes/. [Accessed 16 11
         2019].

[67]   E. Thomas, "In the battle against deepfakes, AI is being
         pitted against AI," Wired, 25 11 2019. [Online]. Available:
         https://www.wired.co.uk/article/deepfakes-ai. [Accessed
         24 01 2020].

[68]   M. Bickert, "Enforcing Against Manipulated Media,"
        Facebook Newsroom, 06 01 2020. [Online]. Available:
        https://about.fb.com/news/2020/01/enforcing-against-
        manipulated-media/. [Accessed 24 01 2020].

[69]   Ctrl Shift Face, "YouTube," Ctrl Shift Face, 08 07 2019.
        [Online]. Available:
        https://www.youtube.com/watch?v=HG_NZpkttXE.
        [Accessed 16 11 2019].

[70]   E. Zakharov, A. Shysheya and E. Burkov, "YouTube,"
        Samsung Research, 21 05 2019. [Online]. Available:
        https://www.youtube.com/watch?feature=youtu.be&v=p
        1b5aiTrGzY. [Accessed 16 11 2019].

[71]   Lovot, "Emotional Robotics™ was developed to make the
        human power to love, even stronger.," Lovot, [Online].
        Available: https://lovot.life/en/technology/. [Accessed 24
        11 2019].

[72]   J. D. Rey, "How robots are transforming Amazon
        warehouse jobs — for better and worse," Vox Recode,
        11 12 2019. [Online]. Available:
        https://www.vox.com/recode/2019/12/11/20982652/rob
        ots-amazon-warehouse-jobs-automation. [Accessed 24
        01 2020].

[73]   M. Simon, "Inside the Amazon Warehouse Where
        Humans and Machines Become One," Wired, 05 06
        2019. [Online]. Available:

https://www.wired.com/story/amazon-warehouse-
robots/. [Accessed 24 01 2020].

[74]   J. Dastin, "Exclusive: Amazon rolls out machines that
pack orders and replace jobs," Reuters, 13 05 2019.
[Online]. Available: https://www.reuters.com/article/us-
amazon-com-automation-exclusive/exclusive-amazon-
rolls-out-machines-that-pack-orders-and-replace-jobs-
idUSKCN1SJ0X1. [Accessed 24 01 2020].

[75]   N. Wingfield, "As Amazon Pushes Forward With Robots,
Workers Find New Roles," The New York Times, 10 09
2017. [Online]. Available:
https://www.nytimes.com/2017/09/10/technology/amaz
on-robots-workers.html. [Accessed 17 11 2019].

[76]   N. Statt, "Amazon says fully automated shipping
warehouses are at least a decade away," The Verge, 01
05 2019. [Online]. Available:
https://www.theverge.com/2019/5/1/18526092/amazon-
warehouse-robotics-automation-ai-10-years-away.
[Accessed 17 11 2019].

[77]   K. Cheung, "Alibaba Launches 'AI Copywriter',"
Algorithm-Xlab, 22 08 2019. [Online]. Available:
https://algorithmxlab.com/blog/alibaba-launches-ai-
copywriter-2/. [Accessed 17 11 2019].

[78]   Z. Wang and L. Theis, "Speedy Neural Networks for
Smart Auto-Cropping of Images," Twitter, 24 01 2018.
[Online]. Available:

https://blog.twitter.com/engineering/en_us/topics/infras
tructure/2018/Smart-Auto-Cropping-of-Images.html.
[Accessed 17 11 2019].

[79]   A. Pasick, "The magic that makes Spotify's Discover
         Weekly playlists so damn good," Quartz, 21 12 2015.
         [Online]. Available: https://qz.com/571007/the-magic-
         that-makes-spotifys-discover-weekly-playlists-so-damn-
         good/. [Accessed 17 11 2019].

[80]   D. Etherington, "Domino's serves up self-driving pizza
         delivery pilot in Houston," Tech Crunch, 17 06 2019.
         [Online]. Available:
         https://techcrunch.com/2019/06/17/dominos-serves-up-
         self-driving-pizza-delivery-pilot-in-houston/. [Accessed
         17 11 2019].

[81]   Domino's, "Domino's Anyware," Domino's, [Online].
         Available: https://anyware.dominos.com/. [Accessed 17
         11 2019].

[82]   Domino's, "DOM Pizza Checker," Domino's, [Online].
         Available: https://dompizzachecker.dominos.com.au/.
         [Accessed 17 11 2019].

[83]   BBC News, "McDonald's uses AI for ordering at drive-
         throughs," BBC, 11 09 2019. [Online]. Available:
         https://www.bbc.com/news/technology-49664633.
         [Accessed 17 11 2019].

[84]   5thru, "The Drive-Thru Reimagined," 5thru, [Online].
        Available: https://www.5thru.com/. [Accessed 17 11
        2019].

[85]   CaliBurger, "Meet Flippy, our new kitchen assistant,"
        CaliBurger, [Online]. Available:
        https://caliburger.com/flippy. [Accessed 17 11 2019].

[86]   J. Graham, "Hamburger-making robot Flippy is back at
        Calif. chain," USA Today, 30 05 2018. [Online]. Available:
        https://eu.usatoday.com/story/tech/talkingtech/2018/05
        /28/hamburger-making-robot-flippy-back-serving-300-
        burgers-day/649370002/. [Accessed 17 11 2019].

[87]   C. Albrecht, "CaliBurger Adds a Second Flippy Robot to
        Make French Fries," The Spoon, 02 10 2019. [Online].
        Available: https://thespoon.tech/caliburger-adds-a-
        second-flippy-robot-to-make-french-fries/. [Accessed 17
        11 2019].

[88]   Samsung, "Get a Glimpse of the Next-generation
        Innovations on Display at Samsung's Technology
        Showcase," Samsung Newsroom, 20 02 2019. [Online].
        Available: https://news.samsung.com/global/get-a-
        glimpse-of-the-next-generation-innovations-on-display-
        at-samsungs-technology-showcase. [Accessed 17 11
        2019].

[89]   Hopper, "Relax — booking travel just got easy," Hopper,
        [Online]. Available: https://www.hopper.com/. [Accessed
        17 11 2019].

[90]   A. Carr, "The Messy Business Of Reinventing Happiness,"
       Fast Company, 15 04 2015. [Online]. Available:
       https://www.fastcompany.com/3044283/the-messy-
       business-of-reinventing-happiness. [Accessed 17 11
       2019].

[91]   A. Mosseri, "Our Commitment to Lead the Fight Against
       Online Bullying," Instagram, 08 07 2019. [Online].
       Available: https://instagram-
       press.com/blog/2019/07/08/our-commitment-to-lead-
       the-fight-against-online-bullying/. [Accessed 17 11 2019].

[92]   Infervision, "About Infervision," Infervision, [Online].
       Available: https://www.infervision.com/site/en.html.
       [Accessed 17 11 2019].

[93]   R. Liao, "China's Infervision is helping 280 hospitals
       worldwide detect cancers from images," Tech Crunch, 30
       11 2018. [Online]. Available:
       https://techcrunch.com/2018/11/30/infervision-medical-
       imaging-280-hospitals/. [Accessed 17 11 2019].

[94]   Babylon, "Babylon understands symptoms you enter
       and provides you with relevant health and triage
       information," Babylon, [Online]. Available:
       https://www.babylonhealth.com/product/ask-babylon.
       [Accessed 17 11 2019].

[95]   P. Schwab and W. Karlen, "PhoneMD: Learning to
       Diagnose Parkinson's Disease from Smartphone Data,"
       Institute of Robotics and Intelligent Systems, 14 11 2018.

[Online]. Available: https://arxiv.org/abs/1810.01485. [Accessed 17 11 2019].

[96] Woebot, "Hi, I'm Woebot," Woebot, [Online]. Available: https://woebot.io/. [Accessed 17 11 2019].

[97] Wysa, "Meet Wysa," Wysa, [Online]. Available: https://www.wysa.io/. [Accessed 17 11 2019].

[98] L. Plummer, "This is how Netflix's top-secret recommendation system works," Wired, 22 08 2017. [Online]. Available: https://www.wired.co.uk/article/how-do-netflixs-algorithms-work-machine-learning-helps-to-predict-what-viewers-will-like. [Accessed 17 11 2019].

[99] M. Kassner, "AI stops identity fraud before it occurs," Tech Republic, 14 01 2016. [Online]. Available: https://www.techrepublic.com/article/ai-stops-identity-fraud-before-it-occurs/. [Accessed 17 11 2019].

[100] E. Guzun, "FIM Launches Nordic AI-Powered Fund," Hedge Nordic, 07 11 2017. [Online]. Available: https://hedgenordic.com/2017/11/fim-launches-first-nordic-ai-powered-fund/. [Accessed 17 11 2019].

[101] E. Guzun, "No One-Trick Pony," Hedge Nordic, 30 11 2017. [Online]. Available: https://hedgenordic.com/2017/11/no-one-trick-pony/. [Accessed 17 11 2019].

[102] E. Guzun, "New AI-Powered Fund Ready to Launch," Hedge Nordic, 27 03 2018. [Online]. Available:

https://hedgenordic.com/2018/03/new-ai-powered-fund-ready-to-launch/. [Accessed 17 11 2019].

[103] Xbox Wired Staff, "Forza Horizon 2: What's a Drivatar, and Why Should I Care?," Xbox Wire, 30 09 2014. [Online]. Available: https://news.xbox.com/en-us/2014/09/30/games-forza-horizon-2-drivatars/. [Accessed 19 11 2019].

[104] Salesforce, "Work smarter with artificial intelligence that's built right into Salesforce.," Salesforce, [Online]. Available: https://www.salesforce.com/products/einstein/overview/. [Accessed 23 11 2019].

[105] Q. V. Le and M. Schuster, "A Neural Network for Machine Translation, at Production Scale," Google AI Blog, 27 09 2016. [Online]. Available: https://ai.googleblog.com/2016/09/a-neural-network-for-machine.html. [Accessed 23 11 2019].

[106] B. Marr, "The Incredible Ways John Deere Is Using Artificial Intelligence To Transform Farming," Forbes, 09 03 2018. [Online]. Available: https://www.forbes.com/sites/bernardmarr/2018/03/09/the-incredible-ways-john-deere-is-using-artificial-intelligence-to-transform-farming/. [Accessed 23 11 2019].

[107] John Deere, "Only John Deere seamlessly connects machines, people, technology, and insights to give you

an advantage.," John Deere, [Online]. Available: https://www.deere.com/en/technology-products/precision-ag-technology/. [Accessed 23 11 2019].

[108]   Viso.ai, "The first cloud platform that helps businesses building their own AI vision solutions to disrupt their industry.," Viso.ai, [Online]. Available: https://viso.ai/#solutions. [Accessed 23 11 2019].

[109]   L. Dormehl, "Self-driving apple harvesting robot suctions the fruit off trees," Digital Trends, 01 04 2019. [Online]. Available: https://www.digitaltrends.com/cool-tech/apple-suctioning-robot-new-zealand-orchard/. [Accessed 23 11 2019].

[110]   S. Ghaffary, "How to avoid a dystopian future of facial recognition in law enforcement," Vox, 10 12 2019. [Online]. Available: https://www.vox.com/recode/2019/12/10/20996085/ai-facial-recognition-police-law-enforcement-regulation. [Accessed 03 01 2020].

[111]   K. Hao, "Police across the US are training crime-predicting AIs on falsified data," Technology Review, 13 02 2019. [Online]. Available: https://www.technologyreview.com/s/612957/predictive-policing-algorithms-ai-crime-dirty-data/. [Accessed 03 01 2020].

[112]   L. Vaas, "Report: Use of AI surveillance is growing around the world," Naked Security, 20 09 2019. [Online]. Available: https://nakedsecurity.sophos.com/2019/09/20/report-use-of-ai-surveillance-is-growing-around-the-world/. [Accessed 23 11 2019].

[113]   N. Statt, "Orlando police once again ditch Amazon's facial recognition software," The Verge, 18 07 2019. [Online]. Available: https://www.theverge.com/2019/7/18/20700072/amazon-rekognition-pilot-program-orlando-florida-law-enforcement-ended. [Accessed 23 11 2019].

[114]   M. Hayes, "New Orleans Police Claim Not To Use Facial Recognition Tech. Emails Reveal That's Not Totally True.," OneZero, 26 08 2019. [Online]. Available: https://onezero.medium.com/new-orleans-police-claim-not-to-use-facial-recognition-tech-emails-reveal-thats-not-totally-true-465f8cd9a71c. [Accessed 23 11 2019].

[115]   R. Haridy, "California officially bans police use of facial recognition in body cameras," New Atlas, 10 10 2019. [Online]. Available: https://newatlas.com/computers/california-bans-police-facial-recognition-body-cameras/. [Accessed 23 11 2019].

[116]   N. Broekhuijsen, "Dutch Police Now Using AI-Powered Surveillance to Catch Drivers on Their Phone," Tom's Hardware, 09 10 2019. [Online]. Available:

https://www.tomshardware.com/news/dutch-police-machine-learning-ai-catches-drivers-on-phone,40596.html. [Accessed 23 11 2019].

[117]   BBC News, "AI cameras to catch texting Australian drivers," BBC, 02 12 2019. [Online]. Available: https://www.bbc.com/news/technology-50630763. [Accessed 07 12 2019].

[118]   D. Fuscaldo, "Hong Kong Police Have AI Facial Recognition Software," Interesting Engineering, 23 10 2019. [Online]. Available: https://interestingengineering.com/hong-kong-police-have-ai-facial-recognition-software. [Accessed 23 11 2019].

[119]   C. Hughes, "British court rules for police use of facial recognition technology," UPI, 04 07 2019. [Online]. Available: https://www.upi.com/Top_News/World-News/2019/09/04/British-court-rules-for-police-use-of-facial-recognition-technology/2851567605583/. [Accessed 23 11 2019].

[120]   M. Hoy, "Police Use of Facial Recognition Tech Approved in Sweden," Bloomberg Law, 25 10 2019. [Online]. Available: https://news.bloomberglaw.com/privacy-and-data-security/police-use-of-facial-recognition-tech-approved-in-sweden. [Accessed 23 11 2019].

[121]   D. Cameron, "Amazon Is Marketing Face Recognition to Police Departments Partnered With Ring: Report,"

Gizmodo, 15 10 2019. [Online]. Available: https://gizmodo.com/amazon-is-marketing-face-recognition-to-police-departme-1839073749. [Accessed 23 11 2019].

[122] An Amazon Employee, "I'm an Amazon Employee. My Company Shouldn't Sell Facial Recognition Tech to Police.," Medium, 16 10 2018. [Online]. Available: https://medium.com/@amazon_employee/im-an-amazon-employee-my-company-shouldn-t-sell-facial-recognition-tech-to-police-36b5fde934ac. [Accessed 23 11 2019].

[123] J. D. Rey, "Jeff Bezos says Amazon is writing its own facial recognition laws to pitch to lawmakers," Vox, 26 09 2019. [Online]. Available: https://www.vox.com/recode/2019/9/25/20884427/jeff-bezos-amazon-facial-recognition-draft-legislation-regulation-rekognition. [Accessed 09 01 2020].

[124] S. Harrison, "Poll Finds Americans Trust Police Use of Facial Recognition," Wired, 09 05 2019. [Online]. Available: https://www.wired.com/story/poll-americans-trust-police-facial-recognition/. [Accessed 23 11 2019].

[125] S. Elks, "Most Britons oppose police use of facial recognition tech: poll," Global News, 24 10 2019. [Online]. Available: https://globalnews.ca/news/6079748/facial-recognition-police-britain/. [Accessed 24 11 2019].

[126] BBC News, "Police officers raise concerns about 'biased' AI data," BBC, 16 09 2019. [Online]. Available: https://www.bbc.com/news/technology-49717378. [Accessed 03 01 2020].

[127] D. Lu, "UK police are using AI to spot spikes in Brexit-related hate crimes," New Scientist, 28 08 2019. [Online]. Available: https://www.newscientist.com/article/mg24332453-500-uk-police-are-using-ai-to-spot-spikes-in-brexit-related-hate-crimes/. [Accessed 24 11 2019].

[128] Reuters, "Amazon ditched AI recruiting tool that favored men for technical jobs," The Guardian, 11 10 2018. [Online]. Available: https://www.theguardian.com/technology/2018/oct/10/amazon-hiring-ai-gender-bias-recruiting-engine.

[129] C. Ledbetter, "Controversial Photo-Editing App Under Fire For Makeup Removal Feature," Huffpost, 15 11 2017. [Online]. Available: https://www.huffpost.com/entry/makeapp-makeup-removal-app_n_5a0c56bde4b0b17ffce1aca1?guccounter=2. [Accessed 24 11 2019].

[130] J. Kastrenakes, "Controversial deepfake app DeepNude shuts down hours after being exposed," The Verge, 27 06 2019. [Online]. Available: https://www.theverge.com/2019/6/27/18761496/deepnu

de-shuts-down-deepfake-nude-ai-app-women.
[Accessed 24 11 2019].

[131] M. Lefkowitz, " AI speeds effort to protect endangered elephants," Cornell Chronicle, 27 08 2018. [Online]. Available: https://news.cornell.edu/stories/2018/08/ai-speeds-effort-protect-endangered-elephants. [Accessed 24 11 2019].

[132] Wildbook, "Wildbook in 90 Seconds," Wildbook, [Online]. Available: http://wildbook.org/doku.php. [Accessed 24 11 2019].

[133] Radar, "Radar: Using the latest AI tools to dynamically create high quality content at massive scale," Radar, [Online]. Available: https://pa.media/radar/. [Accessed 24 11 2019].

[134] Cogito, "Emotional Intelligence and Analysis for Customer Service and Sales," Cogito, [Online]. Available: https://www.cogitocorp.com/product/. [Accessed 24 11 2019].

[135] Google, "Google Nest Learning Thermostat," Google, [Online]. Available: https://store.google.com/us/product/nest_learning_thermostat_3rd_gen?hl=en-US. [Accessed 24 11 2019].

[136] B. J. Born-stein, R. Castan and T. A. Estlin, "Autonomous Exploration for Gathering Increased Science," NASA, 09 2010. [Online]. Available:

https://ntrs.nasa.gov/archive/nasa/casi.ntrs.nasa.gov/20
100033547.pdf. [Accessed 24 11 2019].

[137]   B. Bosker, "Facebook Buys Facial Recognition Firm
Face.com: What It Wants With Your Face," Huffpost, 19
06 2012. [Online]. Available:
https://www.huffpost.com/entry/facebook-buys-face-
com_n_1608996. [Accessed 24 11 2019].

[138]   R. Medway, "Lexus Europe Creates World's Most
Intuitive Car Ad with IBM Watson," IBM, 19 11 2018.
[Online]. Available:
https://www.ibm.com/blogs/think/2018/11/lexus-
europe-creates-worlds-most-intuitive-car-ad-with-ibm-
watson/. [Accessed 24 11 2019].

[139]   Resident Evil Seamless HD Project, "Resident Evil 2
Seamless HD Project," RESHDP, [Online]. Available:
https://www.reshdp.com/re2/. [Accessed 24 11 2019].

[140]   T. Dimson, "Emojineering Part 1: Machine Learning for
Emoji Trends," Instagram Engineering, 01 05 2015.
[Online]. Available: https://instagram-
engineering.com/emojineering-part-1-machine-learning-
for-emoji-trendsmachine-learning-for-emoji-trends-
7f5f9cb979ad#.aevw6iz16. [Accessed 24 11 2019].

[141]   J. Constine, "Snapchat Acquires Looksery To Power Its
Animated Lenses," TechCrunch, 15 09 2015. [Online].
Available: https://techcrunch.com/2015/09/15/snapchat-
looksery/. [Accessed 24 11 2019].

[142]   K. Hill, "How Target Figured Out A Teen Girl Was
        Pregnant Before Her Father Did," Forbes, 16 02 2012.
        [Online]. Available:
        https://www.forbes.com/sites/kashmirhill/2012/02/16/h
        ow-target-figured-out-a-teen-girl-was-pregnant-before-
        her-father-did/. [Accessed 24 11 2019].

[143]   K. Chung, "Generating Recommendations at Amazon
        Scale with Apache Spark and Amazon DSSTNE," AWS Big
        Data Blog, 09 07 2016. [Online]. Available:
        https://aws.amazon.com/blogs/big-data/generating-
        recommendations-at-amazon-scale-with-apache-spark-
        and-amazon-dsstne/. [Accessed 24 11 2019].

[144]   D. Patel, "Site Planning using Location Data," Locale, 17
        10 2019. [Online]. Available: https://medium.com/locale-
        ai/site-planning-using-location-data-ae7814973521.
        [Accessed 24 11 2019].

[145]   S. H. Somanchi, "The mail you want, not the spam you
        don't," Official Gmail Blog, 09 07 2015. [Online].
        Available: https://gmail.googleblog.com/2015/07/the-
        mail-you-want-not-spam-you-dont.html. [Accessed 24 11
        2019].

[146]   Microsoft Azure, "Azure Kinect DK," Microsoft, [Online].
        Available: https://azure.microsoft.com/en-
        gb/services/kinect-dk/. [Accessed 24 11 2019].

[147]   R. Perper, "Chinese government forces people to scan
        their face before they can use internet as surveillance

efforts mount," Business Insider, 02 12 2019. [Online]. Available: https://www.businessinsider.com/china-to-require-facial-id-for-internet-and-mobile-services-2019-10?r=US&IR=T. [Accessed 24 01 2020].

[148]   A. Ma, "China has started ranking citizens with a creepy 'social credit' system — here's what you can do wrong, and the embarrassing, demeaning ways they can punish you," Business Insider, 29 10 2018. [Online]. Available: https://www.businessinsider.com/china-social-credit-system-punishments-and-rewards-explained-2018-4?r=US&IR=T#1-banning-you-from-flying-or-getting-the-train-1. [Accessed 24 11 2019].

[149]   L. Kuo, "China bans 23m from buying travel tickets as part of 'social credit' system," The Guardian, 01 03 2019. [Online]. Available: https://www.theguardian.com/world/2019/mar/01/china-bans-23m-discredited-citizens-from-buying-travel-tickets-social-credit-system. [Accessed 24 11 2019].

[150]   SenseTime, "About Us," SenseTime, [Online]. Available: https://www.sensetime.com/en/about. [Accessed 06 01 2020].

[151]   CNA Insider, "Curbing Unsafe Bus Driver Habits In China, With Artificial Intelligence," CNA Insider, 02 03 2019. [Online]. Available: https://www.youtube.com/watch?v=9F-w3MvQ2qk. [Accessed 07 12 2019].

[152] Fortem Technologies, "Security Elevated™," Fortem
      Technologies, [Online]. Available:
      https://fortemtech.com/. [Accessed 30 11 2019].

[153] M. Brocchetto, C. Dominguez and J. Sterling,
      "Venezuelan President survives apparent drone
      assassination attempt," CNN, 04 08 2018. [Online].
      Available:
      https://edition.cnn.com/2018/08/04/americas/venezuela
      -maduro/index.html. [Accessed 30 11 2019].

[154] Skyline, "Real Estate Investment Meets Artificial
      Intelligence," Skyline, [Online]. Available:
      https://www.skyline.ai/. [Accessed 30 11 2019].

[155] Food and Beverage, "Coca-Cola is Using AI to Put Some
      Fizz in Its Vending Machines," Food and Beverage, 2018.
      [Online]. Available:
      https://foodandbeverage.wbresearch.com/blog/coca-
      cola-artificial-intelligence-ai-omnichannel-strategy.
      [Accessed 30 11 2019].

[156] Adobe Communications Team, "Adobe Research and UC
      Berkeley: Detecting Facial Manipulations in Adobe
      Photoshop," Adobe Blog, 14 06 2019. [Online]. Available:
      https://theblog.adobe.com/adobe-research-and-uc-
      berkeley-detecting-facial-manipulations-in-adobe-
      photoshop/. [Accessed 30 11 2019].

[157]   One Concern, "What we do," One Concern, [Online].
         Available: https://www.oneconcern.com/product/.
         [Accessed 30 11 2019].

[158]   One Concern, "What we believe," One Concern, [Online].
         Available: https://www.oneconcern.com/mission.
         [Accessed 30 11 2019].

[159]   Twentybn, "Meet millie: The AI coach powered by
         TwentyBN technology," Twentybn, [Online]. Available:
         https://20bn.com/. [Accessed 07 12 2019].

[160]   Obvious, "Edmond de Belamy," Obvious, [Online].
         Available: https://obvious-art.com/edmond-de-
         belamy.html. [Accessed 07 12 2019].

[161]   J. Vincent, "How three French students used borrowed
         code to put the first AI portrait in Christie's," The Verge,
         23 10 2018. [Online]. Available:
         https://www.theverge.com/2018/10/23/18013190/ai-art-
         portrait-auction-christies-belamy-obvious-robbie-barrat-
         gans. [Accessed 07 12 2019].

[162]   A. Fabiani, "Robots were used to judge this year's World
         Artistic Gymnastics Championships," Screen Shot, 16 10
         2019. [Online]. Available: https://screenshot-
         magazine.com/the-future/robots-judge-gymnastics/.
         [Accessed 07 12 2019].

[163]   T. Cao, "Meet Fujitsu's AI Gymnastics Judges," Synced
         Review, 26 01 2019. [Online]. Available:

https://medium.com/syncedreview/meet-fujitsus-ai-gymnastics-judges-8cb52613b2a. [Accessed 07 12 2019].

[164] G. Nott, "AI judges readied for Tokyo 2020 gymnastics competitions," CIO, 20 05 2019. [Online]. Available: https://www.cio.com.au/article/669055/jamie-smith-new-cio-hollard-insurance/. [Accessed 07 12 2019].

[165] H. Peterson, "Walmart reveals it's tracking checkout theft with AI-powered cameras in 1,000 stores," Business Insider, 20 06 2019. [Online]. Available: https://www.businessinsider.com/walmart-tracks-theft-with-computer-vision-1000-stores-2019-6?r=US&IR=T. [Accessed 03 01 2020].

[166] Walmart, "AI and the Future of Retail," Walmart, [Online]. Available: https://corporate.walmart.com/IRL/. [Accessed 03 01 2020].

[167] A. Silver, "Introducing Visual Studio IntelliCode," Microsoft, 07 05 2018. [Online]. Available: https://devblogs.microsoft.com/visualstudio/introducing-visual-studio-intellicode/. [Accessed 08 12 2019].

[168] A. Silver, "Re-imagining developer productivity with AI-assisted tools," Microsoft, 04 11 2019. [Online]. Available: https://devblogs.microsoft.com/visualstudio/ai-assisted-developer-tools/. [Accessed 08 12 2019].

[169]   Kite, "Code Faster in Python with Intelligent Snippets,"
        Kite, [Online]. Available: https://kite.com/. [Accessed 08
        12 2019].

[170]   Codota, "AI completions for your Java IDE," Codota,
        [Online]. Available: https://www.codota.com/. [Accessed
        08 12 2019].

[171]   Microsoft, "Sketch2Code," Microsoft, [Online]. Available:
        https://sketch2code.azurewebsites.net/. [Accessed 08 12
        2019].

[172]   Uizard, "Explore the research and technology that
        makes Uizard possible.," Uizard, [Online]. Available:
        https://uizard.io/research/#pix2code. [Accessed 08 12
        2019].

[173]   Microsoft, "Microsoft Program Synthesis using Examples
        SDK," Microsoft , [Online]. Available:
        https://microsoft.github.io/prose/. [Accessed 08 12
        2019].

[174]   S. Gulwani, "Programming by Examples," Databricks,
        [Online]. Available:
        https://databricks.com/session/programming-by-
        examples. [Accessed 08 12 2019].

[175]   DeepCode, "Write better code with the knowledge of the
        global development community," DeepCode, [Online].
        Available: https://www.deepcode.ai/. [Accessed 08 12
        2019].

[176]   AccessiBe, "AI-Powered Web Accessibility: Automatic.
         Simple. Affordable.," AccessiBe, [Online]. Available:
         https://accessibe.com/. [Accessed 08 12 2019].

[177]   Accenture, "The power to seize opportunity before it
         knocks," Accenture, 2016. [Online]. Available:
         https://www.accenture.com/us-en/service-ai-for-
         business-transformation?src=GHG. [Accessed 02 01
         2020].

[178]   J. Schrittwieser, I. Antonoglou, T. Hubert, K. Simonyan, L.
         Sifre, S. Schmitt, A. Guez, E. Lockhart, D. Hassabis, T.
         Graepel, T. Lillicrap and D. Silver, "Mastering Atari, Go,
         Chess and Shogi by Planning with a Learned Model,"
         Cornell University, 19 11 2019. [Online]. Available:
         https://arxiv.org/abs/1911.08265. [Accessed 10 01 2020].

[179]   Planet Money, "< #806: Walmart's Pickle," NPR, 17 11
         2017. [Online]. Available:
         https://www.npr.org/transcripts/564963483?storyId=564
         963483?storyId=564963483&t=1578044504397.
         [Accessed 03 01 2020].

Notes

www.ingramcontent.com/pod-product-compliance
Lightning Source LLC
Chambersburg PA
CBHW021404210526
45463CB00001B/212